Raspberry Pi Pico DIY Workshop

Build exciting projects in home automation, personal health, gardening, and citizen science

Sai Yamanoor

Srihari Yamanoor

BIRMINGHAM—MUMBAI

Raspberry Pi Pico DIY Workshop

Group Product Manager: Rahul Nair

Publishing Product Manager: Rahul Nair

Senior Editor: Shazeen Iqbal

Content Development Editor: Romy Dias

Technical Editor: Rajat Sharma

Copy Editor: Safis Editing

Project Coordinator: Ashwin Kharwa

Proofreader: Safis Editing

Indexer: Subalakshmi Govindhan

Production Designer: Joshua Misquitta

Marketing Coordinator: Sanjana Gupta

First published: May 2022

Production reference: 1100522

Published by Packt Publishing Ltd.

Livery Place

35 Livery Street

Birmingham

B3 2PB, UK.

ISBN 978-1-80181-481-2

www.packt.com

To all the makers who start several projects at a time and dream of finishing them one day.

– Sai Yamanoor

To the wonderful cats that have owned my heart, Squeaky, Bob, Saxon, Gi-Ve, Fluffy Tux, Angel, Grey Cat, and now, Alphie, Fish-Bone, and Saxi!

– Srihari Yamanoor

Contributors

About the authors

Sai Yamanoor is a senior IoT applications engineer at an industrial gases company in Buffalo, NY. He has over 10 years of experience as an embedded systems expert, working on both hardware and software design and implementations. He is a co-author of two books on the use of Raspberry Pi to execute DIY projects, and he has also presented a personal health dashboard at Maker Faires across the country. Sai is currently working on projects aimed at improving the **Quality of Life (QoL)** for people with chronic health conditions.

I want to thank my parents and my brother and co-author, Sri, for all the help and encouragement. I would also like to thank our technical reviewers, Salman Faris and Jonathan Witts, for their insightful comments and for reviewing our work carefully. I would like to thank Rahul Nair for giving us this opportunity to work with Packt. I would also like to thank Romy Dias and Vaidehi Sawant for their patience and for supporting our work.

Srihari Yamanoor is a mechanical engineer with experience spanning medical device design, CAD/CAM, mechatronics, and sustainability. In collaboration with his brother, he develops open source hardware products aimed at public education and awareness. He has multiple certifications in quality assurance, CAD, and FEA. Besides design, manufacturing, and quality, his current interests include behavioral change and working toward active improvement in the fight against diabetes, innovation paradigms and methodologies, and the impact of AI on healthcare. He is the co-author of two books on Raspberry Pi applications and writes blogs on various topics.

I want to thank my parents, mentors, friends, cats, and my brother and co-author, Sai, for all the help and encouragement. Specifically, I would like to thank my mentors Anna Tamura and Dr. Sudhi Gautam. Similarly, I owe gratitude to my friend Satyakanth Thyagaraja for standing by me during tough times. I would also like to thank the Packt team for all their support with this book, and our efforts throughout the years.

About the reviewers

Salman Faris is a digital fabrication and rapid product prototype enthusiast from India who holds a bachelor's degree in computer science and a digital fabrication diploma from the Fab Academy. He is currently working as a technical support engineer at Nebra and is a key contributor to the MakerGram Maker community, where he tinkers with electronics and hardware product development.

Salman is also part of the Edge Impulse expert group, Qubitro, RAK, and Seeed Studio as well as an ambassador and core member of India's largest Maker gathering, Maker Faire Hyderabad, and co-organizer of Maker Fest Kerala.

I'd like to thank Allah first, for His almighty guidance in whatever decisions I make. I'd also like to thank Packt Publishing for the opportunity to review this wonderful book, especially Shagun and Ashwin who managed the review and helped me with guidance and support throughout the process. Thanks also to my parents, siblings, relatives, friends, team Nebra, and mentors.

Jon Witts has worked in IT within the education sector for over 17 years. He holds degrees in fine art and the design and development of e-learning. In his current role, as director of digital strategy, Jon leads all technological solutions in his school, as well as teaching computer science to students aged 11-16. Jon also runs the Hull Raspberry Jam events in his hometown: free coding events for young people using the Raspberry Pi computer. Jon has reviewed a number of titles for Packt and has written his own book, *Wearable-Tech Projects with the Raspberry Pi Zero*, also published by Packt. In his free time, Jon enjoys creating generative art using the p5.js library and incorporating elements from Raspberry Pi physical computing.

I would like to thank my wife, Sally, and our three daughters, Mabel, Ember, and Ada, for all of their support in allowing me to work on this book. I would also like to thank the authors and all the team at Packt for allowing me to be involved in the process of creating this great publication.

Table of Contents

2

Serial Interfaces and Applications

3

Home Automation Projects

6

Designing a Giant Seven-Segment Display

7

Designing a Visual Aid for Tracking Air Quality

Section 3: Advanced Topics

8
Building Wireless Nodes

9
Let's Build a Robot!

10
Designing TinyML Applications

11
Let's Build a Product!

12
Best Practices for Working with the Pico

Preface

As soon as the announcement for the Raspberry Pi Pico was made by the Raspberry Pi Foundation in January 2021, we were excited by the new possibilities that the USD 4 developer board would open for hobbyists, generalists, citizen scientists, professionals, scientists, teachers, and students around the world. Available in various forms, the powerful yet inexpensive microcontroller can indeed stand on its own and work with other tools to help people develop very powerful and elegant solutions. We expect that just like the prior generations of products from the Raspberry Pi Foundation, the Raspberry Pi Pico is going to create another revolution in the realms of technology, education, entertainment, and other societal endeavors.

Based on our experience in penning books and articles on the Raspberry Pi **Single-Board Computers (SBCs)**, we designed this book to dive into new and repeat projects so we may cater to the varying needs of the target audiences, such as students, teachers, engineers, scientists, artists, and tech enthusiasts who want to develop embedded systems that drive cost-effective automation, IoT, robotics, medical devices, and art projects.

We have tried to retain variety in the projects, while also introducing different sensors, programming, interfaces, and other factors sufficient to pave the way for both beginners and advanced readers to ideate and implement projects based on the Raspberry Pi Pico.

Who this book is for

As mentioned, we have developed the materials and projects to cater to a wide variety of readers. You may be a seasoned hobbyist or professional interested in how the Pico can help you with your projects. You may have a little or a lot of experience with electronics, SBCs, microcontrollers, or programming. You may have all the skillsets and be on the hunt for new projects, to entertain yourself or teach in your classroom. This book is aimed at people with a wide variety of backgrounds and experiences.

That said, some basic experience in programming, electronics, and related areas will be very useful in getting through the materials and projects in the book. If you wish to jumpstart your experience with Python programming, you can refer to our other publication, *Python Programming with Raspberry Pi*, also published by Packt.

What this book covers

Chapter 1, Getting Started with the Raspberry Pi Pico, lays out the fundamentals of the Raspberry Pi Pico, the various forms it comes in, the accessories, and how to program the Pico. We also show you how to complete a classic Hello World example and make an LED blink.

Chapter 2, Serial Interfaces and Applications, is the chapter where we explore how to use the serial interfaces of the Raspberry Pi Pico to communicate with sensors, displays, and other hardware. We also demonstrate how to interface a Wi-Fi module and connect the Raspberry Pi Pico to the internet.

Chapter 3, Home Automation Projects, continues with simple home automation projects that can be completed in a weekend, expanding on applications with serial interfaces. We also introduce and demonstrate the Arduino RP2040 Connect and how it can be used in place of a Pico.

Chapter 4, Fun with Gardening, allows us to dig deeper into project implementations with the Pico. We interface a soil sensor to a live plant, measure temperature and soil humidity, and upload the data to an IoT analytics platform and visualize the collected data.

Chapter 5, Building a Weather Station, is a special treat for weather geeks and citizen scientists. We will build a weather station, exploring different sensors and interface options with the Raspberry Pi Pico.

Chapter 6, Designing a Giant Seven-Segment Display, is all about creating great visual aids. We discuss driving the display via the serial port or from within a local network.

Chapter 7, Designing a Visual Aid for Tracking Air Quality, continuing from the previous chapter, is where we demonstrate a different visual aid application, this time taking two different approaches: one using existing data sources and the other using a carbon dioxide sensor to determine air quality.

Chapter 8, Building Wireless Nodes, is where we go beyond Wi-Fi and explore other ways of collecting and transmitting data wirelessly, using LoRa, Sigfox, and Bluetooth. This will allow you to freely develop wireless applications with the Pico.

Chapter 9, Let's Build a Robot!, is for the robotics enthusiasts. We demonstrate how a robot can be controlled with the Pico. In this chapter, we introduce MicroPython as well, for those who plan to keep the code light.

Chapter 10, Designing TinyML Applications, is a gateway for those of you looking to develop AI applications with the Pico. We introduce you to TinyML, a framework specifically focused on lightweight AI applications. We lead you through a keyword recognition example, which will help set the stage for you to take the examples further.

Chapter 11, Let's Build a Product!, takes you on a journey to turbocharge things further and build a product. We demonstrate a method to build a carrier PCB for the Pico, and also how to use a cellular module for connectivity.

Chapter 12, Best Practices for Working with the Pico, is where we close out the book, with tips and tricks that can take your projects with the Pico further. We discuss how the Pico firmware can be updated, how the Arduino IDE can be used to program the Pico, power profiling the Pico, and programming the PIOs.

We hope that the chapters and projects will prepare you for your future adventures with the Raspberry Pi Pico.

To get the most out of this book

Software/hardware covered in the book	Operating system requirements
CircuitPython	Windows, macOS, or Linux
MicroPython	Windows, macOS, or Linux
Arduino IDE (C/C++)	Windows, macOS, or Linux

The projects discussed in this chapter are hardware intensive. In order to maintain consistency, we used the Raspberry Pi Pico across all chapters. You will also need an ESP-32 wireless pack for network connectivity. Each chapter has a list of recommended hardware and we have listed alternatives wherever possible. We leave it up to you to substitute components as you see fit.

If you are using the digital version of this book, we advise you to type the code yourself or access the code from the book's GitHub repository (a link is available in the next section). Doing so will help you avoid any potential errors related to the copying and pasting of code.

If you find any specific hardware-related issues with the code samples shared in the repository, please feel free to create a GitHub issue.

Download the example code files

You can download the example code files for this book from GitHub at `https://github.com/PacktPublishing/Raspberry-Pi-Pico-DIY-Workshop`. If there's an update to the code, it will be updated in the GitHub repository.

We also have other code bundles from our rich catalog of books and videos available at `https://github.com/PacktPublishing/`. Check them out!

Code in Action

The Code in Action videos for this book can be viewed at `https://bit.ly/3OZJb5Z`.

Download the color images

We also provide a PDF file that has color images of the screenshots and diagrams used in this book. You can download it here: `https://static.packt-cdn.com/downloads/9781801814812_ColorImages.pdf`.

Conventions used

There are a number of text conventions used throughout this book.

`Code in text`: Indicates code words in text, database table names, folder names, filenames, file extensions, pathnames, dummy URLs, user input, and Twitter handles. Here is an example: "The `board` module contains definitions of the pins and peripherals specific to the board."

A block of code is set as follows:

```
from machine import Pin
import utime
led = Pin(25, Pin.OUT)
while True:
```

When we wish to draw your attention to a particular part of a code block, the relevant lines or items are set in bold:

```
while True:
    led.toggle()
    utime.sleep(1)
```

Any command-line input or output is written as follows:

```
>>> print("Hello World")
```

Bold: Indicates a new term, an important word, or words that you see onscreen. For instance, words in menus or dialog boxes appear in **bold**. Here is an example: "An impulse can be created from the **Create impulse** tab on the left."

> **Tips or Important Notes**
> Appear like this.

Get in touch

Feedback from our readers is always welcome.

General feedback: If you have questions about any aspect of this book, email us at customercare@packtpub.com and mention the book title in the subject of your message.

Errata: Although we have taken every care to ensure the accuracy of our content, mistakes do happen. If you have found a mistake in this book, we would be grateful if you would report this to us. Please visit www.packtpub.com/support/errata and fill in the form.

Piracy: If you come across any illegal copies of our works in any form on the internet, we would be grateful if you would provide us with the location address or website name. Please contact us at copyright@packt.com with a link to the material.

If you are interested in becoming an author: If there is a topic that you have expertise in and you are interested in either writing or contributing to a book, please visit authors.packtpub.com.

Share your thoughts

Once you've read *Raspberry Pi Pico DIY Workshop*, we'd love to hear your thoughts!
Scan the QR code below to go straight to the Amazon review page for this book and share
your feedback.

https://packt.link/r/1801814813

Your review is important to us and the tech community and will help us make sure we're
delivering excellent quality content.

Section 1:
An Introduction to the Pico

The objective of this section is to introduce the Raspberry Pi Pico, its variants, and the peripherals available on the Raspberry Pi Pico. This part begins slowly by blinking an LED and reviewing the serial interfaces available on the RP2040 microcontroller. Then, we will progress by working on simple home automation and gardening projects.

This section contains the following chapters:

- *Chapter 1, Getting Started with the Raspberry Pi Pico*
- *Chapter 2, Serial Interfaces and Applications*
- *Chapter 3, Home Automation Projects*
- *Chapter 4, Fun with Gardening!*

1
Getting Started with the Raspberry Pi Pico

In this chapter, we would like to delve into a quick introduction to the **Raspberry Pi Pico** and the **RP2040 microcontroller**. We will discuss the Raspberry Pi Pico's features, the RP2040's peripherals, the add-on hardware for the Pico, and development boards for the RP2040 developed by other makers. We will also discuss the programming language options available for the Pico and supplement the chapter by discussing a simple *"Hello World"* example where we print something to the screen and blink a **light-emitting diode** (**LED**).

By the end of this first chapter, you will have gotten started with the Pico and will be ready to start programming the RP2040 microcontroller and start planning to implement projects from the later chapters of this book, as well as thinking ahead to how you can tackle your own projects with the Raspberry Pi Pico!

We are going to cover the following main topics:

- Introducing the Raspberry Pi Pico and RP2040
- Discussing variants of the Pico board
- Soldering the Pico's headers
- Implementing the "Hello World!" example
- Implementing the LED-blinking example
- Identifying useful add-on hardware for the Pico

Technical requirements

The hardware and software required for this introductory chapter will be used throughout the book. In further chapters, we will provide any additional or chapter-specific requirements.

The hardware requirements are listed as follows:

- A laptop or a Raspberry Pi with a **Universal Serial Bus** (**USB**) port
- *Optional*: Soldering equipment including iron, solder, safety glasses, and miscellaneous equipment
- *Optional*: Prototyping breadboard and a jumper wire kit

Code in Action videos for this chapter can be viewed at `https://bit.ly/3MRdYjx`.

Introducing the Raspberry Pi Pico and RP2040

The Raspberry Pi Pico is the latest educational and industrial tool introduced by the Raspberry Pi Foundation. The Pico, a low-cost microcontroller, costs USD 4, and even at the low price point, the Pico packs quite a punch. The Pico is centered on the RP2040, a dual-core **Cortex-M0+** microcontroller. The board comes with a total of 40 pins, where there are 20 pins on each side, as shown in the following screenshot. The Pico also comes with 2 MB of onboard flash memory and an LED on the GP25 (**GP** refers to **General Purpose** Input/Output) button.

Figure 1.1 – Raspberry Pi Pico

The datasheet for the Raspberry Pi Pico is available from here: `https://bit.ly/3cwvlIc`. In this book, we will be making use of the different peripherals available on the Pico in the projects discussed in each chapter. Hence, it is handy to print the pinout provided by the Raspberry Pi foundation (source: `https://bit.ly/3wa0nwq`). This pinout sheet can help with pin selection during project planning. A screenshot of the pinout from Adafruit Industries is shown next. You can purchase them for USD 0.50 from their website.

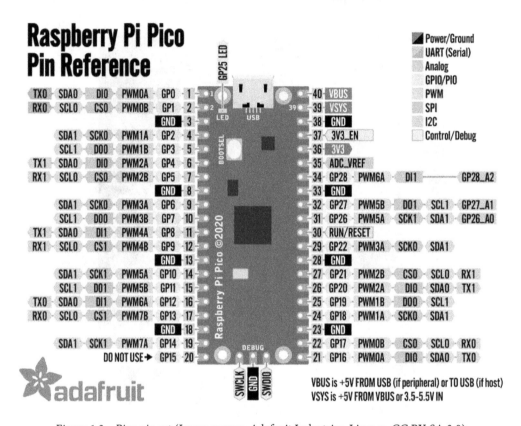

Figure 1.2 – Pico pinout (Image source: Adafruit Industries; License: CC BY-SA 3.0)

The Pico board can be used in various applications involving robots, remote monitoring, citizen science, and so on. In this book, we will walk you through different application examples while exploring the peripherals of the RP2040 microcontroller.

RP2040 microcontroller

The RP2040 is a dual-core ARM Cortex-M0+ microcontroller with 264 **kilobytes (KB)** of **static random-access memory (SRAM)** but does not have have in-built flash memory. The RP2040 comes with a volley of peripherals including **Inter-Integrated Circuit (I2C)**, **Serial Peripheral Interface (SPI)**, and **Programmable Input/Output (PIO)**. The PIO on the RP2040 microcontroller enables you to design your own interface, such as an additional **universal asynchronous receiver-transmitter (UART)** interface or a video interface. In *Chapter 12, Best Practices for Working with the Pico*, we will discuss using the PIO peripheral.

Here is a list of the resources for RP2040:

- The datasheet for the RP2040 is available at the following link:

  `https://bit.ly/3rw41x5`

- The datasheet for the Raspberry Pi Pico is available at the following link:

  `https://bit.ly/3cwv1Ic`

- A video from the Raspberry Pi foundation on the RP2040's PIO can be found at the following link:

  `https://bit.ly/39ni6Xg`

- Resources for the RP2040 from the Raspberry Pi Foundation can be found by visiting the following link:

  `https://bit.ly/3flFLv9`

We recommend that you download Pico's datasheet along with the RP2040 datasheet. It will come in handy as a reference during development, and we will refer you to the datasheet at certain points in this book for more information.

Discussing variants of the Pico board

Since the launch of the Raspberry Pi Pico, there have been several developer board variants that include the RP2040 from various open hardware companies. These are boards that come with the RP2040 microcontroller and they are outlined in more detail here:

- **SparkFun Thing Plus – RP2040** (USD 17.95): This is an open source development board from SparkFun (`https://bit.ly/2NS5vUn`). The **Thing Plus** comes in the **Feather form factor** from Adafruit. Something unique about this board is that it comes with a microSD card holder and an individually addressable RGB LED. If you are not familiar with the Feather form factor, it simplifies prototyping due to its stacking capability and the ecosystem of prototyping tools available in the Feather form factor. In the following screenshot, you can notice the top (on the right) and bottom (on the left) sides of the RP2040 Thing Plus board:

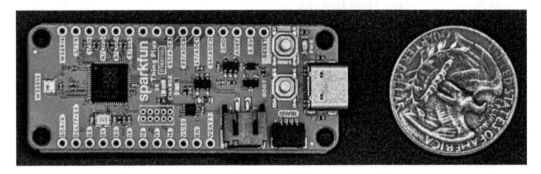

Figure 1.3 – SparkFun Thing Plus (RP2040)

- **SparkFun MicroMod RP2040 Processor** (USD 11.95): This is another variant from SparkFun (`https://bit.ly/3clp0hG`). It comes with 16 MB of onboard flash memory. It comes in the MicroMod form factor that makes use of the M.2 standard. In the following screenshot, you can find the top and bottom sides of the RP2040 MicroMod board. You will notice a notch in a half-moon shape that is used to fasten the board to a carrier board using an M2.5 screw:

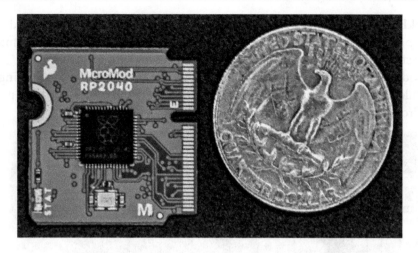

Figure 1.4 – MicroMod RP2040 Processor

SparkFun also makes carrier boards for the MicroMod ecosystem. For example, the carrier board (https://bit.ly/3cnlrHF) shown in the following screenshot was designed to drive a **high-definition multimedia interface** (**HDMI**) display using the RP2040:

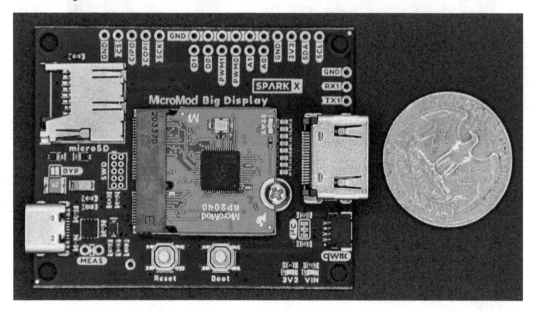

Figure 1.5 – MicroMod Big Display Board for the RP2040 processor

- **SparkFun Pro Micro – RP2040** (USD 9.95): The Pro Micro – RP2040 board (`https://bit.ly/3cnhVgH`) is a variant that belongs to the relatively small ecosystem of the Pro Micro family of boards. It comes with 16 MB of flash, individually addressable RGB LEDs, and castellated pads that enable soldering the module directly onto another **printed circuit board** (**PCB**). The castellated pins of the Pro Micro are shown in the following screenshot:

Figure 1.6 – SparkFun Pro Micro – RP2040

- **Pimoroni Tiny 2040** (USD 11.55): This board from Pimoroni (`https://bit.ly/3d9f7Tf`) is about the size of a quarter and comes with 8 MB of flash and an RGB LED. The castellated pads enable it to be soldered onto your custom PCB directly. We must point out that you will need a cutout to solder the board onto your custom board. This is because the microcontroller in this development board is on the bottom side, as shown in the following screenshot. We will demonstrate using this board on your custom PCB.

Figure 1.7 – Pimoroni Tiny 2040

- **Adafruit Feather RP2040** (USD 11.95): As the name indicates, this board from Adafruit (`https://bit.ly/3cm3tW0`) is a *Feather* board for the RP2040 microcontroller. As with the SparkFun Thing Plus, it packs a punch with a *Qwiic/STEMMA* connector and comes with 8 MB of flash. At the time of writing this book, this board was out of stock. Here's a screenshot showing the board:

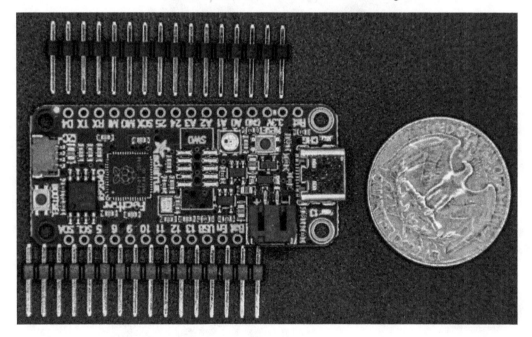

Figure 1.8 – Adafruit Feather RP2040

- **Adafruit ItsyBitsy RP2040** (USD 9.95): This board from Adafruit (`https://bit.ly/3sqdB5R`) is an addition to their *Itsy Bitsy* line of products. In terms of its pinouts, it is identical to other Itsy Bitsy products from Adafruit. This board comes with 8 MB of onboard flash memory. This Itsy Bitsy variant, shown along with the Feather board in the following screenshot, is breadboard-friendly. This enables the board to be embedded into your project:

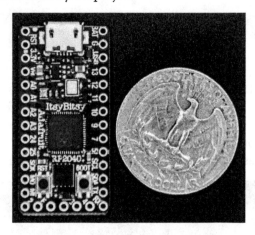

Figure 1.9 – Adafruit ItsyBitsy RP2040

- **Adafruit QT Py RP2040** (USD 9.95): This board (`https://bit.ly/31U2O1q`) is an addition to the *QT Py* family (pronounced "cutie pie") of products from Adafruit. This board also comes with 8 MB of onboard flash memory. The castellated pads of the board shown in the following screenshot enable a PCB to be designed whereby the board could be embedded in your design. Since the RP2040 microcontroller is located on the bottom side, you need to ensure that your design has a cutout to accommodate the QT Py.

Figure 1.10 – Adafruit QT Py RP2040

The variants we discussed here are not comprehensive, but we wanted to present some options on getting started with the RP2040 microcontroller. For example, if you are familiar with the Feather form factor, you could get started with the *Thing Plus* board from *SparkFun* or the *Feather* board from *Adafruit*, discussed in this section. You can use any board of your choice, but the mode of use and interface may differ according to the variant. We will try to highlight any differences wherever possible.

Where to buy the Pico

The Pico costs USD 4 and you can buy it from any Raspberry Pi *distributor*. You can check out the list of Raspberry Pi distributors at this link: `https://bit.ly/3dgra1a`. You can buy variants of the Pico from the links provided with their description.

We must note that the Raspberry Pi Pico was not in stock in the US at the time of writing this chapter. This can be attributed to supply-chain constraints due to the ongoing Covid-19 pandemic at the time. It was also difficult to purchase the variants for the same reason. Things may change in the future.

We must also note that the RP2040 microcontroller was not available for purchase at the time of writing this chapter.

In this section, we discussed variants of the Pico and where to buy them. In the next section, we will take a look at setting up the Pico.

Soldering the Pico's headers

In this section, we will discuss setting up the Pico for our upcoming projects. This includes soldering the headers and an optional **three-dimensional (3D)-printed** enclosure.

Soldering the headers

The Pico comes with 40 pins in two rows of 20 pins. We need to **solder** the headers to access the peripherals of the RP2040 microcontroller for our project.

> **Important Note**
> Soldering the headers requires prior training and adult supervision. Do not attempt soldering without prior training. Here is a tutorial on soldering: `https://bit.ly/3focmjM`.

You can purchase the headers from the same source as the Pico. For example, you could purchase it from the following link: `https://bit.ly/3d9rrUT`. The steps to be carried out for soldering include the following:

1. It is easier to solder the headers with a breadboard. Arrange the headers on a breadboard and stack the Pico on top of it, as shown in the following screenshot:

Figure 1.11 – Pico with headers stacked on a breadboard

2. If you are not quick at soldering the individual pins, you might end up damaging the breadboard due to the excess heat. The following image shows the pins of the Pico soldered:

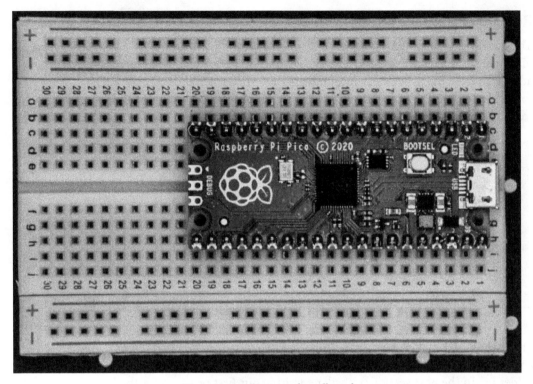

Figure 1.12 – Pico on a breadboard

Now that we have soldered the headers, we are ready to take it for a spin. In the next section, we will review the optional step of adding a reset button to the Pico.

Implementing the "Hello World!" example

In this section, we will discuss a **hello world** example. In any programming language, the first coding exercise is to print *"Hello World"* to the screen. We will discuss this example using **MicroPython** and **CircuitPython**.

Reset button for the Pico (optional)

The Pico does not come with a reset button. In order to reset the Pico, you would have to disconnect and reconnect the USB cable, which can be tedious. This can be overcome by adding a reset button between pin numbers 28 and 30 (note the extra button right next to the microcontroller), as shown in the following image (you can get this button fixture from this link: https://bit.ly/3w8AmO2):

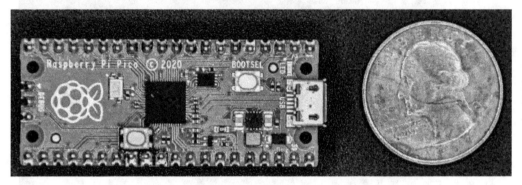

Figure 1.13 – Reset button on the Pico

The reset button makes it easier to restart the program during development. Adding a reset button is optional and it is not recommended if you are not comfortable with soldering. In the next section, we will be writing our first program for the Pico.

MicroPython

We will be using **Thonny IDE** (**IDE** stands for **integrated development environment**) for this example. Thonny IDE can be downloaded from https://thonny.org. The IDE is available for Windows, Mac, and Linux operating systems.

If you use a Raspberry Pi for your programming needs, it comes pre-installed with Thonny. You can launch Thonny from **Menu | Programming** and scroll down the drop-down button to find **Thonny Python IDE**, as illustrated in the following screenshot:

Figure 1.14 – Thonny IDE location on the Raspberry Pi Desktop

Now, let's prepare the Pico by flashing the *MicroPython* binary.

Flashing the MicroPython binary

The Pico needs to be connected to the computer in a specific manner to flash the banner. The steps to flash the binary are outlined here:

1. The first step is to download the binary from `https://bit.ly/31nBMFW`.

2. Press and hold the **BOOTSEL** button shown in the following screenshot as you connect a MicroUSB cable to your Pico. Ensure that the other end of the USB cable is connected to the computer.

Figure 1.15 – Pico BOOTSEL button

3. The Pico should enumerate as a storage device on your computer, which should look like something similar to this:

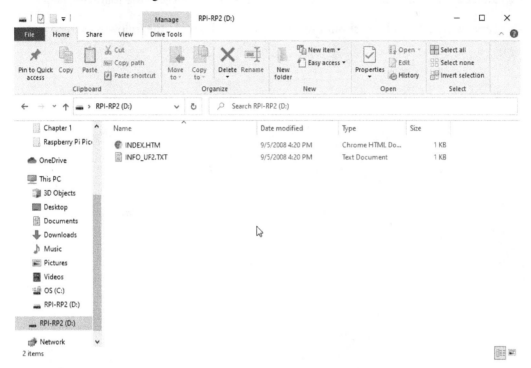

Figure 1.16 – Raspberry Pi Pico enumerated as a storage device

4. Next, copy over the binary onto the storage drive. The Pico will reset itself and we are ready to write programs in MicroPython.

Now, we are ready to write our first program in MicroPython!

Writing our first program

Let's launch Thonny and take it for a test drive! You will need to take the following steps:

1.   Launch Thonny and you should see a window, as shown in the following screenshot:

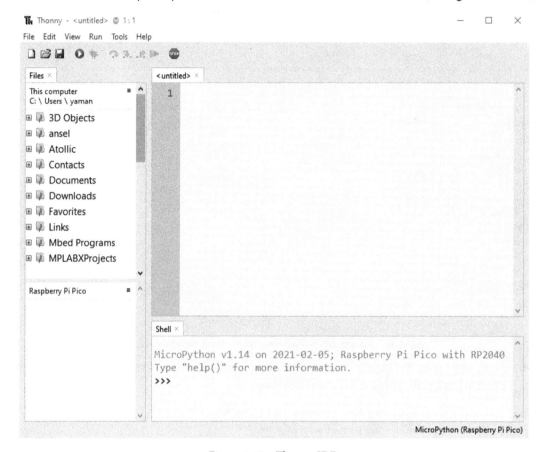

Figure 1.17 – Thonny IDE

2. We are going to be running our first program using the interpreter running on your Pico. If you are not familiar with Python interpreters, they enable the testing of your code as you write them. Go to **Run**, and then click on **Select Interpreter**. From here, you will select the following options, as shown in *Figure 1.18*:

    - **MicroPython (Raspberry Pi Pico)**

    - **< Try to detect port automatically >**

    These options can be seen in the following screenshot:

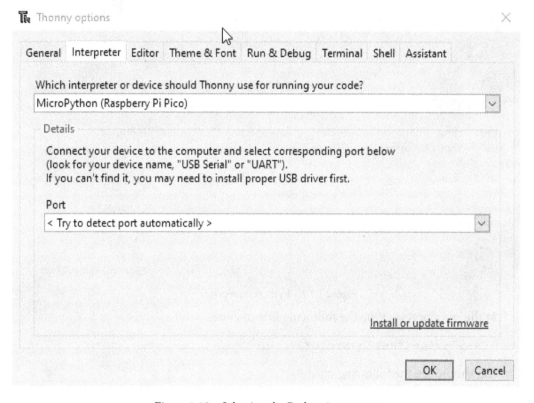

Figure 1.18 – Selecting the Python interpreter

3. Now, you should see the Python interpreter, as shown in the following screenshot. This is the MicroPython interpreter running on the Pico. Let's take it for a test drive.

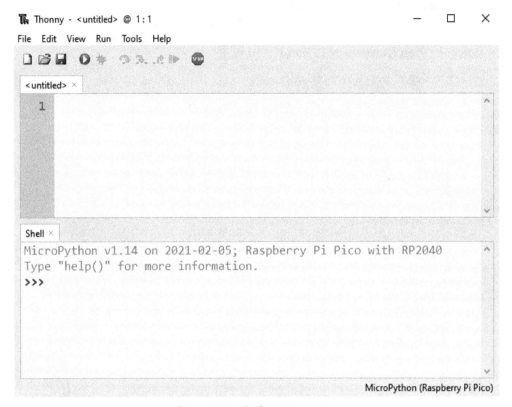

Figure 1.19 – Python interpreter

4. At the prompt >>>, type the following line of code:

```
>>> print("Hello World")
```

You should see the following output:

```
Shell
MicroPython v1.14 on 2021-02-05; Raspberry Pi Pico with RP2040
Type "help()" for more information.
>>> print("Hello World")
  Hello World
>>>
```

Figure 1.20 – Interpreter output

You have just finished writing your first program on your Raspberry Pi Pico! In the next example, we will write a script that starts running continuously when the Pico is powered by a USB cable.

Implementing the LED-blinking example

In the previous section, we used an interpreter to write our program. An interpreter can be helpful when testing code or finding out more information about the modules being imported. In this example, we will discuss writing a script that runs automatically on powering the Pico. We will discuss the *"hello world"* example of getting started with electronics where we will **blink** an **LED** at a 1-second interval.

The Pico comes with a green LED on the board and its location is shown in the following image:

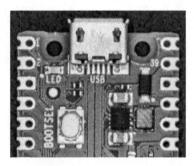

Figure 1.21 – Pico LED location

We will make this LED blink at a 1-second interval—that is, we will turn it on and off for a second. In order to write this program, we will need the `machine` and `utime` modules.

According to the Pico datasheet, the onboard LED is connected to **general-purpose I/O (GPIO)** pin number 25. We will proceed as follows:

- We will use the `Pin` class from the `machine` module to control the LED.
- We will be using the `utime` module to introduce the delay between turning ON and turning OFF the LED.

You will now need to take the following steps:

1. Let's take a look at the following code sample:

    ```
    from machine import Pin
    import utime
    ```

```
led = Pin(25, Pin.OUT)

while True:
    led.toggle()
    utime.sleep(1)
```

> **Exercise**
>
> Develop a practice of actively using an interpreter while you are writing code. In the interpreter, import the `machine` and `utime` modules, try executing `help(machine)`, `help(utime)`, and find out for yourself.

2. Create a new file and enter the preceding code snippet shown in *Step 1*. Set the file location as **Raspberry Pi Pico**, as seen in the following screenshot to the right:

Figure 1.22 – File location

3. Save the file as `main.py` and your code should automatically begin execution.

4. Try disconnecting the USB cable and reconnecting it. You will notice that the script starts running automatically.

5. To stop the code execution, click on the **STOP** button located on the toolbar, as shown in the following screenshot:

```
[ main.py ] ×
1  from machine import Pin
2  import utime
3
4  led = Pin(25, Pin.OUT)
5
6  while True:
7      led.toggle()
8      utime.sleep(1)
```

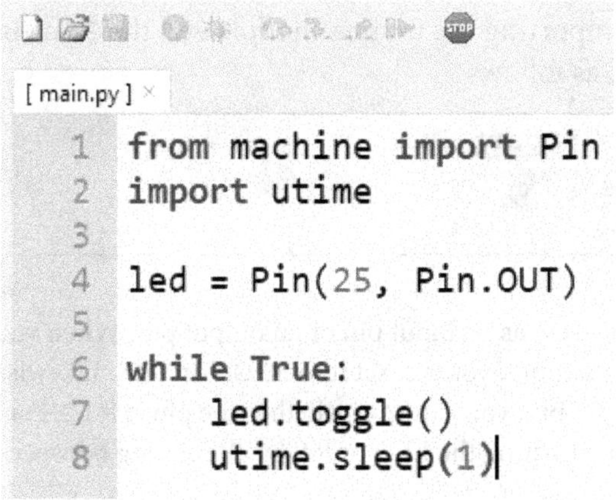

Figure 1.23 – Pressing STOP to interrupt execution

6. To resume execution, click the **Run Current Script** button, which is indicated by the green *play* button indicated at the top of the page, as shown in the following screenshot:

```
[ main.py ] ×
1  from machine import Pin
2  import utime
3
4  led = Pin(25, Pin.OUT)
5
6  while True:
7      led.toggle()
8      utime.sleep(1)
```

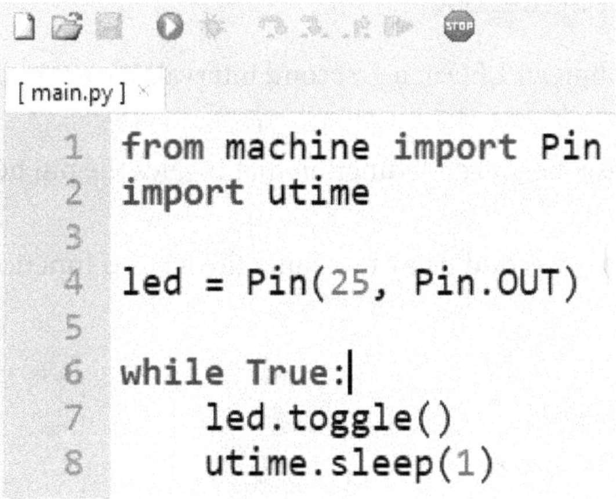

Figure 1.24 – Clicking Run Current Script to resume execution

In the next section, we will take a closer look at the code.

Description of the code sample

In this section, we will discuss in detail the code sample shown previously, as follows:

1. We get started by importing the `utime` module and the `Pin` class from the `machine` module, as follows:

    ```
    from machine import Pin
    import utime
    ```

    > **GPIO**
    >
    > A GPIO pin can be used as an input pin or an output pin. When you use a GPIO pin as an output pin, you can set it to HIGH or LOW. Likewise, when you use it as an input pin, you can read whether the pin is HIGH or LOW. In this example, we are turning the LED on/off by alternating between HIGH and LOW states.

2. Next, we declare the **GPIO** pin 25 as an output pin using an object belonging to the `Pin` class. The first argument in the following code snippet refers to the pin number (pin 25), while the second argument sets the pin as an output pin:

    ```
    led = Pin(25, Pin.OUT)
    ```

3. Now, we need to blink an LED at a 1-second interval. We are going to do this inside an infinite loop.

4. The `led` object has a `toggle()` function that toggles the pin between the ON and OFF states.

5. We will introduce a 1-second delay by calling the `sleep` function after toggling the pin, as follows:

    ```
    while True:
        led.toggle()
        utime.sleep(1)
    ```

Congratulations on your first step with the Raspberry Pi Pico! Now, we will discuss the same example in CircuitPython.

CircuitPython example

In this section, we will discuss using **CircuitPython** for programming the Pico. We will discuss the same LED-blinking example using Pico. We will also install the *Mu IDE* to the program by using CircuitPython.

Flashing the CircuitPython binary

In this section, we will flash the CircuitPython binary onto the Pico. The installation process is the same as for MicroPython. The binary can be downloaded from `https://bit.ly/31pnLI4`.

Once you have downloaded the binary, follow the instructions from the *MicroPython* section.

Coding with Mu

In this section, we will install the Mu IDE. The IDE is available for download from `https://bit.ly/3ruxDKW`. The IDE is available for Windows, Linux, and Mac operating systems.

Raspberry Pi installation

On a Raspberry Pi, Mu can be installed as follows:

1. Go to **Menu** | **Preferences** | **Recommended Software** | **Select Mu** to choose Mu from the list of software to install.

2. Mu can be launched from **Menu** | **Programming** | **Mu**.

In the next section, we will review writing our first program with Mu.

Launching Mu

The steps to writing a program with Mu include the following:

1. Connect the Pico to your computer/Raspberry Pi using a USB cable.

2. Once Mu is launched, we need to set the programming mode. Since we are programming in CircuitPython, we will set it to **CircuitPython**, as shown in the following screenshot:

Figure 1.25 – Selecting programming mode

3. You can also change the programming mode from the main window of the IDE, as shown in the following screenshot:

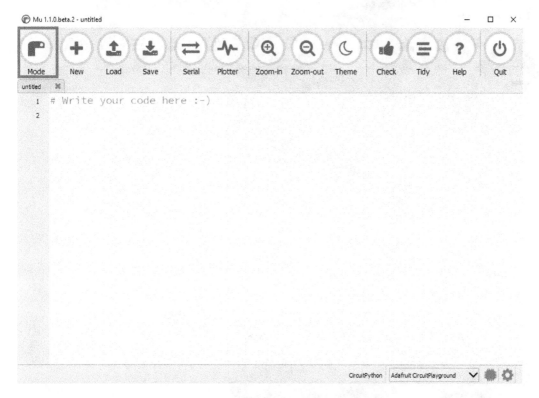

Figure 1.26 – Changing Python mode

Note the Pico being automatically detected in the preceding screenshot.

4. Click the **Serial** button to launch access to the Python interpreter, as shown here:

Figure 1.27 – Serial button location

5. Launch the Python interpreter by pressing *Ctrl + C*, and it should take you to the >>> **REPL** prompt, as shown toward the bottom of the following screenshot:

Figure 1.28 – CircuitPython interpreter on Pico

Finally, in this process, repeat the `Hello World!` example from the *MicroPython* section by using the interpreter.

Second LED-blinking example

We will discuss the same LED-blinking example using CircuitPython and observe the differences between the two flavors of Python. Let's take a quick look at the following code snippet:

```python
import time
import board
import digitalio


led = digitalio.DigitalInOut(board.LED)
```

```
led.direction = digitalio.Direction.OUTPUT

while True:
    led.value = True
    time.sleep(1)
    led.value = False
    time.sleep(2)
```

Let's review the code sample, as follows:

1. We get started by importing the `time`, `board`, and `digitalio` modules. The `time` module is used to introduce a delay between turning the LED ON and OFF.

2. The `board` module contains definitions of the pins and peripherals specific to the board. In this example, we are making use of the `LED` constant from the `board` module to drive the onboard LED on the Raspberry Pi Pico.

3. The `digitalio` module provides access to the Pico's peripherals. In this example, we need to declare the LED pin (GPIO pin 25) as an output pin:

```
led = digitalio.DigitalInOut(board.LED)
led.direction = digitalio.Direction.OUTPUT
```

4. In the first line of the preceding code snippet, we are declaring `led` as an instance of the `DigitalInOut` class.

5. In the second line, we are setting the direction of the `led` pin to be an output pin. We are making use of the `Direction` class from the `digitalio` module.

6. Next, we enter an infinite loop where we turn the LED on/off, as follows:

```
while True:
    led.value = True
    time.sleep(1)
    led.value = False
    time.sleep(2)
```

7. In the first line of the `while` loop, we set the value to be `True`. This turns ON the LED. This is followed by a 1-second delay. This is achieved by calling `time.sleep(1)`.

8. In the third line of the `while` loop, we set the value to be `False`. This turns OFF the LED. This is also followed by a 1-second delay.

9. We want the script to launch upon reset. Load the `code.py` file located on the Pico that is currently enumerated on your computer as a storage device. The **Load** button is located on the top toolbar.

10. Type the code sample we discussed into `code.py` and save it.

11. Press *Ctrl + D* from the CircuitPython interpreter and you should notice the LED blinking on the Pico.

Congratulations on writing your first CircuitPython program for the Raspberry Pi Pico!

CircuitPython or MicroPython?

In this chapter, we discussed examples with both CircuitPython and MicroPython. The examples were somewhat identical and share a similar structure. What are their differences and which flavor of Python should you use for your development?

The short answer is that *it is up to you*. For the sake of consistency, we will be discussing all examples in CircuitPython using the Mu IDE.

Both implementations have a wide user base and libraries for add-on hardware. CircuitPython was spun off MicroPython by Adafruit. CircuitPython can be helpful while using sensor breakout boards from Adafruit Industries.

> **Thonny versus Mu IDE**
>
> In this chapter, you might have noticed that we used Thonny for the MicroPython example and Mu for the CircuitPython example. We wanted to demonstrate the various tools available for the Raspberry Pi Pico. You can even use a simple text editor for your development. We will show you how to save and upload your code to the Pico.

In the next section, we will discuss add-on hardware for the Pico.

Identifying useful add-on hardware for the Pico

In this section, we will discuss the **add-on hardware** available for the Pico. We must note that we are only discussing hardware designed specifically for the Pico. The list is not comprehensive, and we picked examples on the basis of their differences. You are welcome to select any development board of your choosing, but this is not necessary. A simple breadboard should suffice and would be assembled as shown in the following screenshot:

Figure 1.29 – Pico on a breadboard

In the next example, we will take a look at add-on hardware where the Pico sits right next to the breadboard.

Pico Breadboard Kit (USD 19.31)

As the name suggests, this board comes with a breadboard, four LEDs, four pushbuttons, and a buzzer. There are a pair of headers to assemble the Pico onto the board and two rows of headers to access all the pins available on the Pico board. An image of the board can be seen in the following screenshot. This board can be helpful for an absolute beginner in electronics. The board can be purchased from `https://bit.ly/3tV7aIa`.

Figure 1.30 – Pico Breadboard Kit

The Pico breadboard can be helpful for an absolute beginner in electronics.

Pico GPIO Expansion Board (USD 10.34)

This is a development that provides access to all the pins of the Pico. There are two rows of *male* and *female* headers on both sides, as shown in the following screenshot. This board can be purchased from `https://bit.ly/3rprobq`.

Figure 1.31 – Pico GPIO Expansion Board

The two rows of male and female headers enable the use of male and female jumper cables for prototyping.

Pico HAT Expansion (USD 13.79)

This development board enables the interfacing of any Raspberry Pi **HAT** (which stands for **Hardware Attached on Top**) to the Pico. It comes with a 2x20 header that enables a HAT to be stacked on the board and can be seen in the following screenshot. The board also provides access to the pins of the Pico board and is connected to the HAT pinout. The board can be purchased from here: `https://bit.ly/31YRfpu`. The web page also provides pin mapping from the Pico to the HAT.

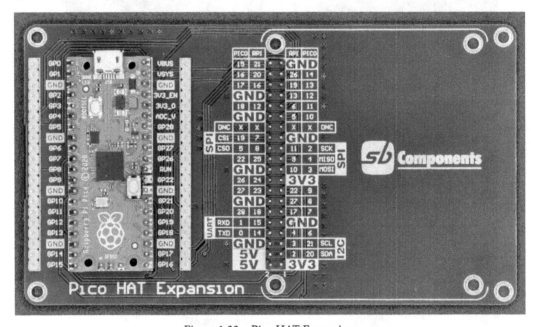

Figure 1.32 – Pico HAT Expansion

This board enables a Pico to be added to your existing Raspberry Pi project. It could also enable you to use your Raspberry Pi HATs with your Pico.

Grove Shield for Pi Pico (USD 3.90)

The **Grove Shield** board helps to connect the Pico to the **Grove ecosystem** from *Seeed Studio*. In case you are not familiar with this, the Grove ecosystem consists of modular boards for prototyping in electronics. As you can see from the following screenshot, the board consists of a series of connectors that enables it to be interfaced to sensors and actuators. The board can be purchased from `https://bit.ly/2NZG3MO`.

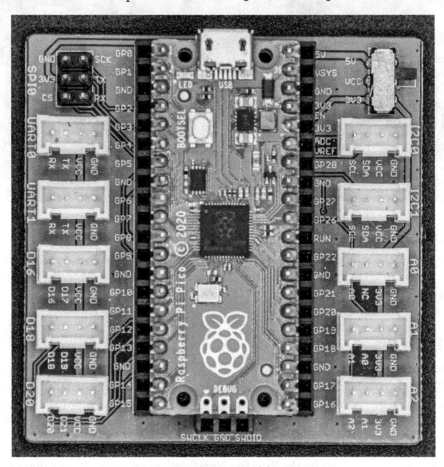

Figure 1.33 – Grove Shield for Pi Pico

In the next example, we will discuss prototyping with stackable hardware.

Pimoroni Pico Decker (Quad Expander) (USD 16.55)

As the name suggests, this board enables the interfacing of up to four expansion boards to the Pico, as shown in the following screenshot, but it is important to ensure that there are no pin conflicts between the add-on boards. The board can be purchased from `https://bit.ly/31XpWMt`:

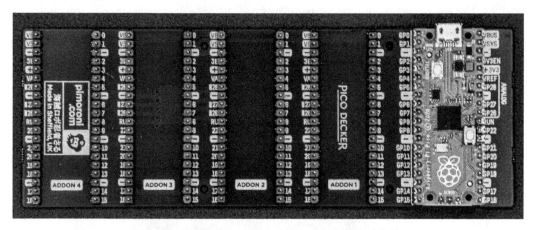

Figure 1.34 – Pimoroni Pico Decker (Quad Expander)

In this section, we reviewed various add-on hardware available for the Raspberry Pi Pico. The list is not comprehensive—for example, *Feather/Thing Plus* boards come with stackable hardware known as **FeatherWings**. Similarly, *MicroMod* boards come with their own ecosystem of add-on hardware.

Summary

In this chapter, we introduced you to the Raspberry Pi Pico. We discussed the peripherals of the RP2040 microcontroller and variants of the Pico development board from other hardware manufacturers. We also discussed the add-on hardware available for the Pico. We soldered the headers for the Pico and added a reset button. We also discussed a *"Hello World!"* example and an LED-blinking example using CircuitPython and MicroPython.

Now that you are done setting up the Pico and have familiarized yourself with the options available to program the Pico, we will discuss the features of the RP2040 microcontroller using practical examples.

Join us in the next chapter, where we will review the communication interfaces available on the Pico!

2
Serial Interfaces and Applications

In this chapter, we are going to discuss the various communication **interfaces** available on the Pico. We are discussing them all in one place because they will come in handy in the projects we will discuss in the upcoming chapters. The topics discussed in this chapter will be useful for interfacing with sensors, driving actuators, displaying messages, and communicating with a **Wi-Fi module** to connect to your local Wi-Fi network to publish and retrieve data from the cloud. This chapter will also help you when it comes to communicating with cellular modules that act like peripheral devices.

If you are familiar with the topics discussed in this chapter, skip right ahead to the next chapter. If not, we recommend playing with the code samples discussed in this chapter to gain a basic understanding of the communication interfaces available on the Pico. In the last section of this chapter, we are going to discuss the setup of an ESP32 wireless pack. We will be using this wireless pack in all the projects discussed in this book.

We are going to cover the following main topics in this chapter:

- Installing the requisite libraries

- Using the UART interface to communicate between two Pico boards

- Interfacing sensors using the I2C interface

- Displaying temperature data using SPI

- Setting up the wireless pack

Technical requirements

The following hardware is used in this chapter:

- 2 x Raspberry Pi Pico (link: `https://bit.ly/3AJtoAf`) – USD 4

- DHT20 temperature and humidity sensor (link: `https://bit.ly/3q5KVAx`) – USD 4.50 *or* HTU21D-F temperature sensor: (link: `https://bit.ly/3gCM51T`) – USD 10.95

- 128x32 SPI OLED display (link: `https://bit.ly/3yAwHJY`) – USD 17.50

- Pico Omnibus – dual expander (link: `https://bit.ly/3sr2GJR`) – USD 10.50

- Pico wireless pack (link: `https://bit.ly/3yPuoT9`) – USD 16.75

> **Component Selection**
> You might have noticed that we recommended two temperature sensors. We wanted to provide alternatives so that you can choose a component according to your budget.

The code samples discussed in this chapter are available for download from here: `https://github.com/PacktPublishing/Raspberry-Pi-Pico-DIY-Workshop/tree/main/chapter_02`.

Code in Action videos for this chapter can be viewed at `https://bit.ly/3w9qPH2`.

In the next section, we will get started by installing the requisite libraries.

Installing requisite libraries

In this section, we will install the requisite libraries for the temperature sensor, the display, and the wireless pack. The libraries are all a part of the Adafruit CircuitPython bundle. The latest bundle can be downloaded as a ZIP file from `https://circuitpython.org/libraries`. We used the bundle version meant for CircuitPython 6.x.x.

After downloading the ZIP file, extract its contents so that we can copy the libraries we need for the project.

> **CircuitPython Installation**
>
> We are assuming that you have installed CircuitPython on your Pico. If you are not familiar with the installation process, we recommend following the installation process from *Chapter 1, Getting Started with the Raspberry Pi Pico.*

HTU21D-F temperature sensor

If you are using the HTU21D-F temperature sensor (shown in the following figure), it requires two libraries, namely `adafruit_bus_device` and `adafruit_htu21d`. From the library folder, copy over the `adafruit_bus_device` folder along with the `adafruit_htu21d.mpy` binary to the `lib` folder of your Pico:

Figure 2.1 – HTU21D-F temperature and humidity sensor

In the next section, we will review the libraries needed for the DHT20 temperature sensor.

DHT20 temperature and humidity sensor

If you are using the DHT20 temperature sensor (shown in the following figure), it requires two libraries, namely `adafruit_bus_device` and `adafruit_ahtx0`. From the library folder, copy over the `adafruit_bus_device` folder along with the `adafruit_ahtx0.mpy` binary to the `lib` folder of your Pico. The temperature sensor chip inside the DHT20 sensor is an **AHT20** temperature sensor. Hence, we are installing the `adafruit_ahtx0` library.

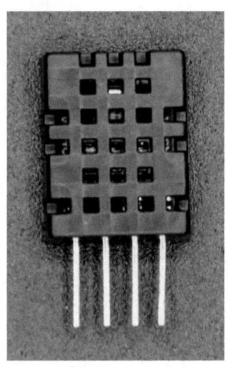

Figure 2.2 – DHT20 temperature and humidity sensor

Next, we will install the libraries for the OLED display.

OLED display (driven by SSD1306)

The OLED display requires three libraries, namely `adafruit_bus_device`, `adafruit_displayio_ssd1306`, and `adafruit_display_text`. Since we installed the first library in the previous section, we will copy over the folder named `adafruit_display_text` and the `adafruit_displayio_ssd1306.mpy` binary from the library bundle to the `lib` folder of your Pico.

Figure 2.3 – 128x32 SPI OLED display

Next, we will install the libraries for the wireless pack.

Wireless pack

We need the `adafruit_esp32spi` library for the wireless pack since we already installed `adafruit_bus_device`. Copy over the folder (with the same name) to the `lib` folder of your Pico. We will also need the `adafruit_reqests.mpy` binary from the bundle.

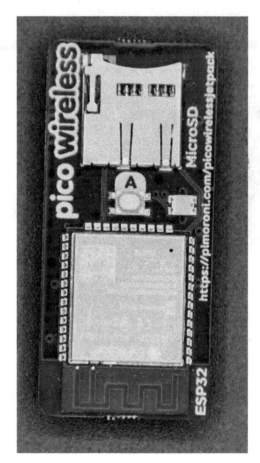

Figure 2.4 – ESP32 wireless pack for Pico

Now that we have installed the requisite libraries, we will get started by testing the UART interface.

Using the UART interface to communicate between two Pico boards

In this section, we will discuss the **Universal Asynchronous Receiver/Transmitter (UART)** interface and its applications. We will be making use of two Pico boards to transmit a message from one Pico to another. The second Pico echoes back the incoming message. Before we jump into the example, let's take a quick look at the UART interface.

The UART interface

The UART interface is a communication protocol using two lines, namely a *transmitter* and a *receiver* (shown in the following figure). As the name suggests, the protocol is asynchronous; that is, there is no reference clock signal for the communication. The communication happens at a preset speed known as a **baud rate**. Typical baud rates include 9600, 19200, 38400, and so on.

The UART interface is used to communicate with sensors such as GPS sensors, dataloggers, and so on. The following figure shows the connection between two devices. The **receiver (RX)** of one device is connected to the **transmitter (TX)** of the other device and their ground pins are tied to each other. In order to initiate communication between them, both devices need to operate at a pre-configured baud rate, data frame size, and so on. Usually, sensors with a UART output, these specifications are mentioned in the datasheet. As an exercise, we suggest reading through the specifications of the following GPS sensor and finding out its baud rate: `https://www.adafruit.com/product/746`.

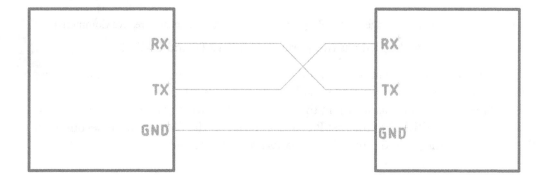

Figure 2.5 – UART communication setup

We recommend the following article for a deep dive on the UART interface: `https://bit.ly/3e9fvD0`.

Setting up the Pico

In this section, we will review wiring up the two Raspberry Pi Picos. We are going to make use of the port **UART0** highlighted in the rectangle of the following figure:

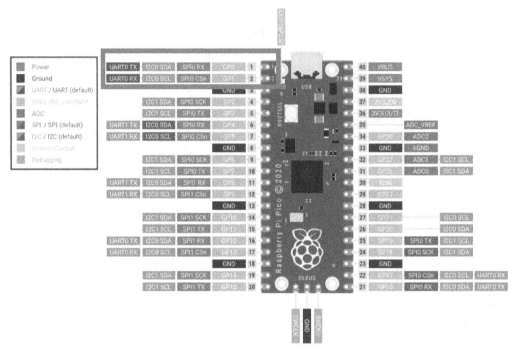

Figure 2.6 – Pico pinout (image source: Raspberry Pi Foundation, License: Creative Commons)

As shown in *Figure 2.6*, the Pico has two ports named **UART0** and **UART1**.

We are going to use UART0 in this example. To transmit messages between the boards, we need to connect the transmitter of the first Pico (Pin 1, GPIO0) to the second pin (Pin 2, GPIO1) of the second Pico as shown in the **Fritzing schematic** in *Figure 2.7*. Similarly, the second pin (Pin 2, GPIO 1) of the first Pico is connected to the first pin of the second Pico (Pin 1, GPIO 0). The ground pins of both boards are tied together.

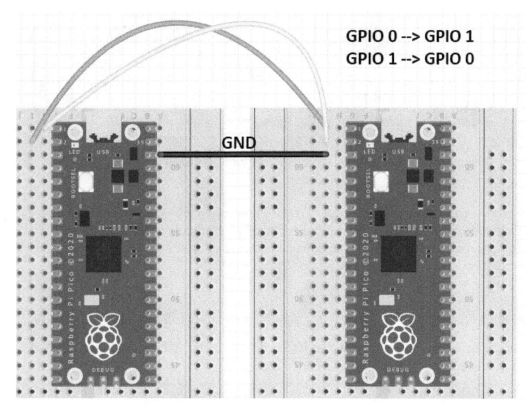

Figure 2.7 – Pico UART schematic (source: Pico part by vanepp – License: Creative Commons)

Now that we have wired up the two Pico boards, let's talk a look at programming them.

Programming the Pico boards

In this example, we are working with two Pico boards, namely a transmitter and a receiver. Let's review the code used in both the Pico boards. While they are similar in their structure, there are certain differences between them due to their function.

Programming the transmitter

The first Pico board captures input from the user via the **serial terminal**. The captured input is then transmitted to the second Pico. Let's take a quick look at the code sample (available for download here: `https://bit.ly/2ZI79gO`):

```
import board
import busio
uart = busio.UART(board.GP0, board.GP1, baudrate=9600)
```

```
while True:
    # capture message
    message = input("Enter a message:").encode("utf-8")
    uart.write(message)
    # receive echo
    bytes_waiting = uart.in_waiting
    if bytes_waiting:
        incoming_msg = uart.read(bytes_waiting)
        print(incoming_msg)
```

Now, let's review the individual blocks of code. The first step is to import the requisite modules into our program. In CircuitPython, the board module contains all the pin definitions required for the Pico. The busio module contains the UART class needed for this example:

```
import board
import busio
```

In the previous section, we discussed *GPIO 0* and *GPIO 1* pins. We will initialize the UART class at a baudrate of 9600 as can be seen here:

```
uart = busio.UART(board.GP0, board.GP1, baudrate=9600)
```

Now, we will enter an infinite loop where we capture incoming messages using the input() function as follows:

```
while True:
    # capture message
    message = input("Enter a message:").encode("utf-8")
```

The string input is converted into bytes using encode. This is because the write() method of the UART class only accepts a byte array:

```
uart.write(message)
```

The next step is to wait for an incoming message from the second board as follows:

```
    # receive echo
    bytes_waiting = uart.in_waiting
    if bytes_waiting:
```

```
incoming_msg = uart.read(bytes_waiting)
print(incoming_msg)
```

The in_waiting property of the UART class returns the number of bytes waiting in the buffer. If there is a message waiting in the buffer, we read it using the read() method of the UART class. The message is printed to the terminal.

Save the code sample as code.py to the first Pico board (called the transmitter) and in the next section, we will review the code sample for the second board (known as the receiver).

Programming the receiver

As mentioned before, the code sample is identical to the previous one. Let's review the program for the second board as follows (available for download here: https://bit.ly/2Y99WiU):

```python
import time
import board
import busio

uart = busio.UART(board.GP0, board.GP1, baudrate=9600)

while True:
    # await incoming message
    bytes_waiting = uart.in_waiting
    if bytes_waiting:
        incoming_msg = uart.readline()
        print(incoming_msg)
        # re-transmit
        uart.write(incoming_msg)
```

Let's review what happens inside the while loop:

1. If there are bytes in the buffer (returned by the in_waiting property), we read it using the readline() method. The method reads until a newline character is reached. We use readline() instead of read() because the incoming message contains a newline character and helps print the entire message captured by the input() function in the previous code sample.

2. Save the program as code.py on the second Pico and it is time to test them.

In the next section, we will test our code using the Mu IDE.

Testing the code

In order to test our code, we will be making use of two instances of the Mu IDE. We need to know the serial port numbers of the Pico. We will demonstrate the test in a Windows environment. The process should be similar in other operating systems.

Let's take the following steps to test the board:

1. Plug in the first Pico board and launch **Device Manager** as shown in *Figure 2.8* (from the **Control Panel**). Identify the serial port number of the first board as shown in the following figure:

> ☐ Network adapters
> ☐ Portable Devices
> ⌄ ☐ Ports (COM & LPT)
>   ☐ USB Serial Device (COM3)
> ☐ Print queues
> ☐ Printers
> ☐ Processors

Figure 2.8 – Note the COM port number from Device Manager

2. Plug in the second Pico board and identify the serial port number (as shown in the following figure):

> ☐ Network adapters
> ☐ Portable Devices
> ⌄ ☐ Ports (COM & LPT)
>   ☐ USB Serial Device (COM3)
>   ☐ USB Serial Device (COM6)
> ☐ Print queues
> ☐ Printers
> ☐ Processors
> ☐ Security devices

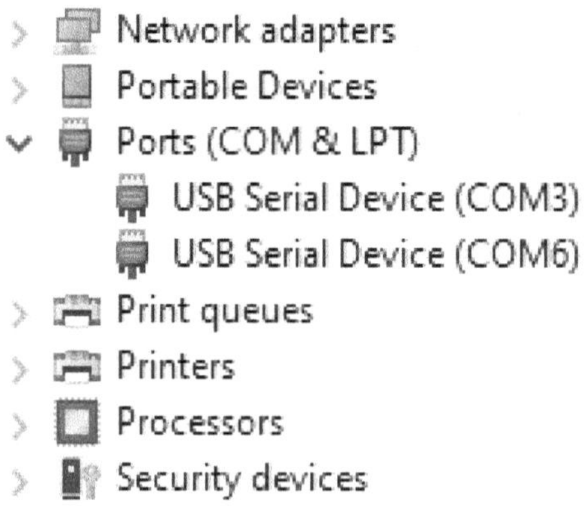

Figure 2.9 – Note the COM port of the second Pico board

3. If you haven't already, launch the first instance of the Mu IDE and ensure that the IDE is set to CircuitPython mode. This can be accomplished with the mode button in the top-left corner of the IDE's toolbar.

4. Now, we need to select the serial port of the first Pico board. In the bottom-right corner of the Mu IDE, there is a drop-down menu of the boards connected to the computer. If you hover over one of the devices, the serial number should be highlighted (as shown in the following figure). Select the serial port of the first board.

Figure 2.10 – Selecting a serial port from the Mu IDE

5. Now, launch a second instance of the Mu IDE and repeat the process for the second Pico board.

6. Open the serial port on both instances of the Mu IDE from the **Serial** button on the toolbar (as shown in the following figure).

Figure 2.11 – Serial button on the toolbar

7. When you open the serial session, you will see a blank window as shown in the following figure. This means that the code is already running. If you see the Python interpreter already, something is wrong with the code.

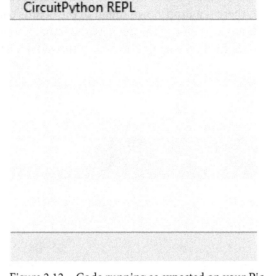

Figure 2.12 – Code running as expected on your Pico

8. Press *Ctrl* + *C* to interrupt execution and *Ctrl* + *D* to reload the code as shown in the following figure:

```
CircuitPython REPL

Traceback (most recent call last):
  File "code.py", line 8, in <module>
KeyboardInterrupt:

Code done running.
Auto-reload is on. Simply save files over USB to run them or enter REPL to disable.

Press any key to enter the REPL. Use CTRL-D to reload.
soft reboot

Auto-reload is on. Simply save files over USB to run them or enter REPL to disable.

code.py output:
Enter a message:
```

Figure 2.13 – Reload the code

9. Repeat the steps for the second Pico board and you should see something like what's shown in the following figure:

```
CircuitPython REPL
KeyboardInterrupt:

Code done running.
Auto-reload is on. Simply save files over USB to run them or enter REPL to disable.

Press any key to enter the REPL. Use CTRL-D to reload.
soft reboot

Auto-reload is on. Simply save files over USB to run them or enter REPL to disable.

code.py output:
```

Figure 2.14 – Output of the second Pico board

10. Enter any message on the first board (transmitter) and you should see it on the
    second (receiver) as shown in the following figure.

```
code.py output:                    code.py output:
Enter a message:Hello              b'Hello'
b'\x00'                            b'Hello'
Enter a message:World              b'\x00'
b'Hello'                           b'Hello'
Enter a message:!                  b'World'
b'World'                           b'!'
Enter a message:|
```

Figure 2.15 – Echo of messages from the second board

11. When you enter a second message on the first Pico (transmitter), you should see an
    echo of the first message transmitted back by the other Pico board (receiver). In the
    previous message, the first message was Hello. It was transmitted to the second
    board and echoed back. When the second message, World, is entered, the message
    echoed back is printed to the screen. This is due to the coding logic. The input()
    function waits until the user input is received and then prints any messages from
    the buffer to the screen.

12. When you are done testing, you can close the session by clicking on the **Serial**
    button on the toolbar.

In the next section, we will discuss some applications of the UART interface.

Applications of the UART interface

We would like to show you some hardware that makes use of the UART interface on the Pico microcontroller, namely the RP2040. The following figure shows a Raspberry Pi Build HAT – an add-on board for the Raspberry Pi. It enables interfacing sensors and controlling motors. The HAT is designed using an RP2040 microcontroller and according to the documentation provided by the Raspberry Pi foundation, it is controlled via the UART interface. Have you come across hardware that can be controlled via UART?

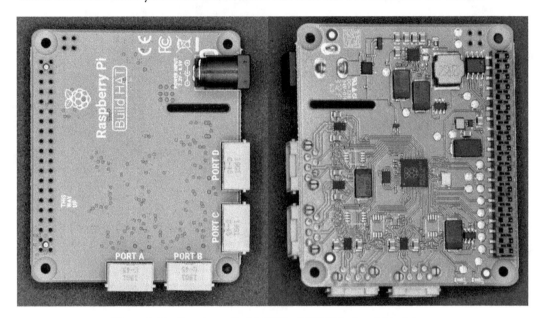

Figure 2.16 – Raspberry Build HAT – RP2040 in the right picture

In the next section, we will discuss the **I2C interface**.

Interfacing sensors using the I2C interface

In this section, we will make use of the I2C (pronounced *I-squared-C*) interface to read temperature and humidity from the *HTU21D* sensor. We will read the temperature and display it on an *OLED display*.

Introduction to the I2C interface

In the previous section, we discussed the UART interface, which is asynchronous; that is, there is no reference clock signal. Now, we are going to discuss the I2C interface, which is synchronous and typically consists of a clock pin (for the reference clock signal) and a data pin. The following figure shows a schematic representation of devices on an I2C bus where we have a host device, which is usually a microcontroller such as the RP2040, and the sensors interface using the clock and data lines. Each sensor on the bus has a unique address that enables the host to communicate with the devices present on the bus. The following figure shows the host and peripheral devices on an I2C bus.

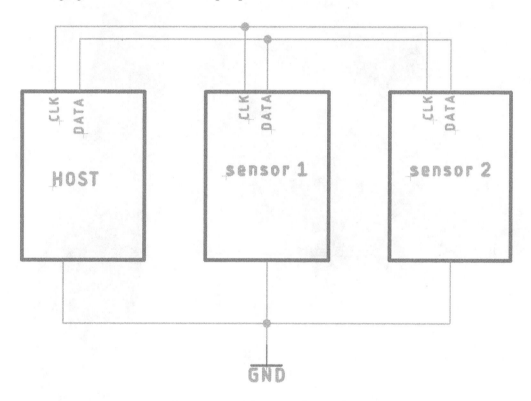

Figure 2.17 – Schematic of an I2C bus

While the UART interface is designed for communication between two devices, the I2C interface is designed to communicate between a host and multiple peripheral devices. For a deep dive, we recommend watching this excellent tutorial on the I2C interface: https://youtu.be/qeJN_8OCiMU.

An I2C interface typically consists of a primary or host device and a secondary or peripheral device. The I2C bus can include multiple host and peripheral devices. An I2C interface is typically used for communicating for sensors, displays, and so on. In the following figure, the Pico is interfaced with a cellular module from `https://blues.io` called the Notecard. The Pico serves as the host device, and it is interfaced to a cellular module that serves as a peripheral. This enables the Pico to collect data from other sensors on the I2C bus and publish it to the cloud. We created a keypad phone that sends you a text message. You can read more about the project here: `https://bit.ly/3pY7J5e`.

Figure 2.18 – Pico interfaced to a peripheral cellular device

In the next section, we will review the need for pull-up resistors on the I2C bus.

Pull-up resistors

The clock and data lines of an I2C bus are open-drain outputs. They operate by driving the circuit low, but they cannot drive it high. Hence, pull-up resistors are needed for normal operation (as shown in the following figure):

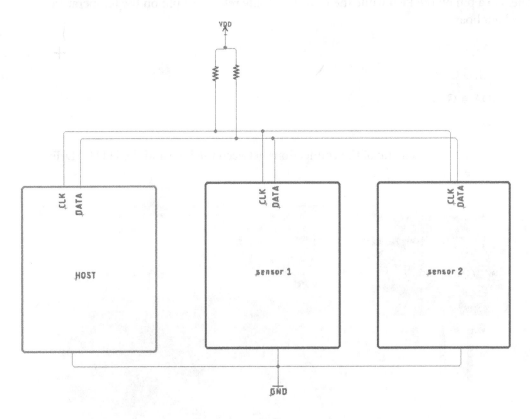

Figure 2.19 – Pull-up resistors

The pull-up resistor values depend on various factors. We found this information from Texas Instruments to be very useful in calculating pull-up resistor values: `https://bit.ly/30dLM7G`. Usually, most I2C prototyping breakout boards come with onboard pull-up resistors. You also might have an option to disconnect them if there is a pull-up resistor elsewhere in the circuit. It is essential to have pull-up resistors on your I2C bus for normal operation.

Now that we have reviewed the I2C interface, we are going to interface a temperature sensor with the Pico.

Testing the HTU21D-F temperature sensor

In this section, we will discuss interfacing the **HTU21D-F** temperature sensor with the Raspberry Pi Pico. The temperature sensor is interfaced with the Pico as shown in the following figure. The connections are as follows, where the left-hand side of the arrow refers to a pin on the Pico while the right-hand side refers to a pin on the temperature breakout board:

- 3.3V → VIN
- SCL → GP9
- SDA → GP8
- GND

Here is a Fritzing schematic of the connections between the Pico and the HTU21D-F temperature sensor.

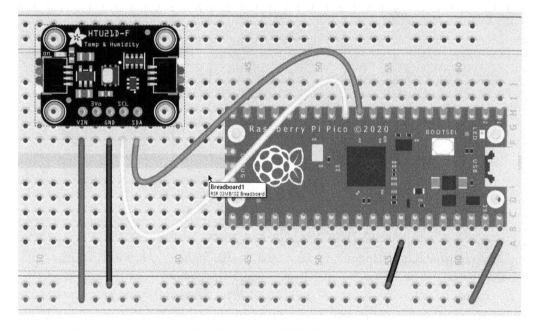

Figure 2.20 – Interfacing the Pico with an HTU21D temperature/humidity sensor

Now let's review the code needed to read temperature and humidity from the HTU21D sensor.

HTU21D-F temperature sensor code

We are assuming that you have installed the requisite libraries for the temperature sensor (from the earlier section). Let's take a quick look at the code required to interface the temperature sensor (available here: `https://bit.ly/3jZD3Nn`).

Once you have installed the library, it is very easy to read the temperature from the sensor. Let's take a quick look at the code sample from Adafruit:

1. The first step is to import the requisite modules:

    ```
    import time
    import board
    import busio
    from adafruit_htu21d import HTU21D
    ```

2. The next step is to initialize the `I2C` class from the `busio` module and the `HTU21D` class:

    ```
    i2c = busio.I2C(board.GP9, board.GP8)
    sensor = HTU21D(i2c)
    ```

3. Now, we can start reading the temperature in a loop:

    ```
    while True:
        print("\nTemperature: %0.1f C" % sensor.temperature)
        print("Humidity: %0.1f %%" % sensor.relative_
    humidity)
        time.sleep(2)
    ```

4. Putting it all together, we have the following:

    ```
    import time
    import busio
    import board
    from adafruit_htu21d import HTU21D
    i2c = busio.I2C(board.GP9, board.GP8)
    sensor = HTU21D(i2c)
    while True:
        print("\n Temperature: %0.1f C" % sensor.temperature)
        print("Humidity: %0.1f %%" % sensor.relative_
    humidity)
        time.sleep(2)
    ```

5. Once we save our code sample to the Pico as `code.py`, we should be able to read the temperature and humidity from the sensor (shown in the following screenshot using the Mu IDE).

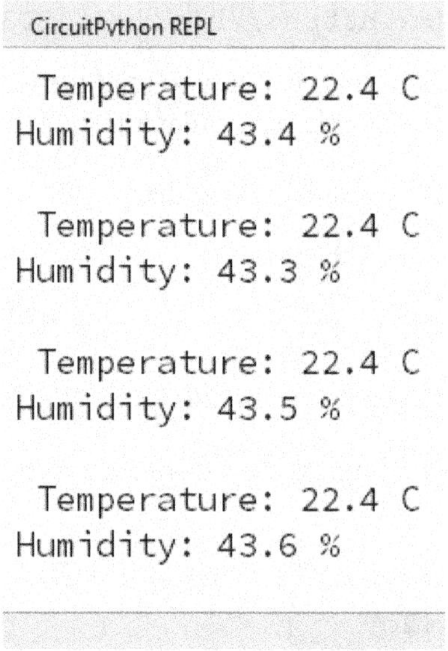

CircuitPython REPL

```
Temperature: 22.4 C
Humidity: 43.4 %

Temperature: 22.4 C
Humidity: 43.3 %

Temperature: 22.4 C
Humidity: 43.5 %

Temperature: 22.4 C
Humidity: 43.6 %
```

Figure 2.21 – Temperature and humidity data from HTU21D

In the next section, we will review testing the AM2320 sensor.

Testing the DHT20 temperature sensor

In this section, we will discuss interfacing the **DHT20** temperature sensor with the Raspberry Pi Pico. The following figure shows the pinouts of the DHT20 sensor (link to datasheet: `https://bit.ly/31GrBQo`).

Pins	Name	Describe
1	VDD	Power supply (2.2v to 5.5v)
2	SDA	Serial data bidirectional port
3	GND	Ground
4	SCL	Serial clock bidirectional port

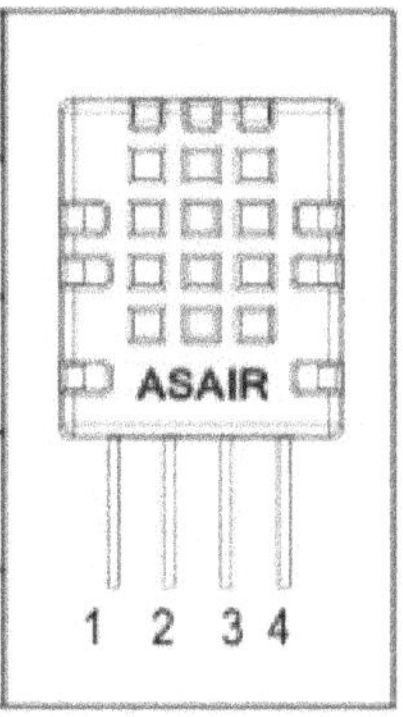

Figure 2.22 – DHT20 sensor pinout

The temperature sensor is interfaced to the Pico as shown in the following figure. The connections are as follows, where the left-hand side of the arrow refers to a pin on the Pico while the right-hand side refers to a pin on the temperature (pinouts shown in the previous figure):

- 3.3V → VIN
- SCL → GP9
- SDA → GP8
- GND

A Fritzing schematic of the connections between the Pico and the AHT20 temperature sensor is shown in the following figure. As mentioned earlier, the DHT20 sensor contains an AHT20 sensor inside. Hence, we created the schematic using an AHT20 part.

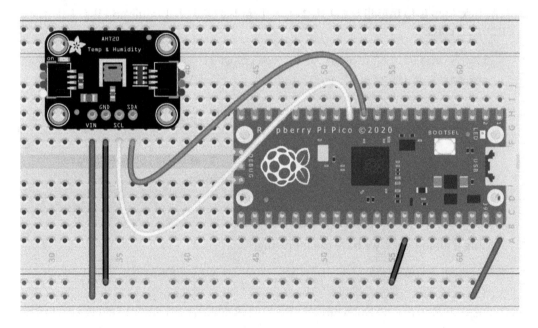

Figure 2.23 – Interfacing the Pico to an AHT20 temperature/humidity sensor

Now let's review the code needed to read temperature and humidity from the AHT20 sensor.

AHT20 temperature sensor code

We are assuming that you have installed the requisite libraries for the temperature sensor (from the earlier section). Let's take a quick look at the code required to interface the temperature sensor.

Once you have installed the library, it is very easy to read the temperature from the sensor. Let's take a quick look at the code sample from Adafrui:.

1. The first step is to import the requisite modules:

    ```
    import time
    import board
    import busio
    from adafruit_ahtx0 import AHTx0
    ```

2. The next step is to initialize the I2C class from the busio module and the HTU21D class:

    ```
    i2c = busio.I2C(board.GP9, board.GP8)
    sensor = AHTx0(i2c)
    ```

3. Now, we can start reading the temperature in a loop:

    ```
    while True:
        print("\nTemperature: %0.1f C" % sensor.temperature)
        print("Humidity: %0.1f %%" % sensor.relative_
    humidity)
        time.sleep(2)
    ```

4. Putting it all together, we have the following:

    ```
    import time
    import busio
    import board
    from adafruit_ahtx0 import AHTx0
    i2c = busio.I2C(board.GP9, board.GP8)
    sensor = AHTx0(i2c)
    while True:
        print("\n Temperature: %0.1f C" % sensor.temperature)
        print("Humidity: %0.1f %%" % sensor.relative_
    humidity)
        time.sleep(2)
    ```

5. Once we save our code sample to the Pico as `code.py`, we should be able to read the temperature and humidity from the sensor (shown in the following screenshot using the Mu IDE).

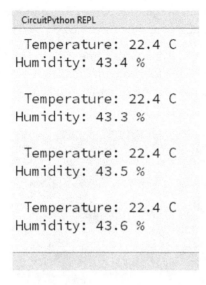

Figure 2.24 – Temperature and humidity data from HTU21D

Now that we have tested the HTU21D temperature sensor, we will take a quick look at the Qwiic/STEMMA ecosystem.

The Qwiic ecosystem

The *Qwiic* ecosystem consists of a series of sensors, displays, and so on, introduced by SparkFun with polarized connectors that enable the easy interface of devices using the I2C interface. You can learn more about the Qwiic interface here: `https://www.sparkfun.com/qwiic`.

While the Qwiic boards from SparkFun operate only at 3.3V, Adafruit Industries has designed boards that are 5V tolerant. Adafruit's ecosystem of boards that make use of the I2C interface is called STEMMA. You can check out the STEMMA ecosystem here: `https://bit.ly/3buXSM3`. Since the Qwiic and STEMMA boards carry the same connector, they are cross-compatible, but the Qwiic boards operate only at 3.3V signal levels. You must be careful while mixing the boards.

The Feather RP2040 board

In the following figure, we have interfaced the STEMMA HTU21D temperature sensor to the Adafruit Feather RP2040 board using a **Qwiic** cable. In case you are not familiar, the RP2040 Feather board is a variant of the Raspberry Pi Pico in the *Adafruit Feather* form factor. You can learn more about it here: `https://bit.ly/3mz5G5t`.

The HTU21D and the Feather RP2040 boards come with a Qwiic connector and they help simplify prototyping as shown in the following figure. You will notice that a simple cable assembly containing polarized connectors on either end is used to interface the sensor to the feather board.

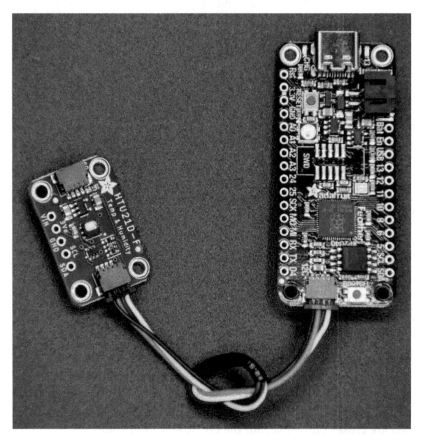

Figure 2.25 – Feather RP2040 interfaced to HTU21D using a Qwiic cable

The Qwiic connectors also enable daisy chaining boards together. In the following figure, you will notice that two temperature sensor prototyping boards are daisy-chained together. This eases prototyping without messy wires on a breadboard and helps avoid any mix-ups encountered while using a breadboard.

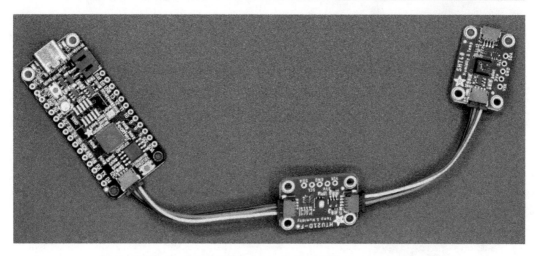

Figure 2.26 – Feather RP2040 daisy-chained with temperature sensor

Familiarity with the UART and I2C interface can help with interfacing displays, controlling motors, and debugging code with information. In the next section, we are going to discuss troubleshooting on the I2C bus.

Troubleshooting

You might run into scenarios where you need to troubleshoot your I2C bus connections on the breadboard. This could be something as simple as mixing up the clock and data lines of the sensor. In CircuitPython, there is a handy scanning tool available to look for devices present on the I2C bus. This helps determine whether the device is connected properly and whether we can identify it. Let's take a quick look at the code available from the CircuitPython documentation (link: https://bit.ly/3H7oUrq). The code sample is available for download along with this chapter as i2c_scanning.py:

1. The first step is to import the requisite modules and initialize the I2C bus:

```
import time
import board
import busio

i2c = busio.I2C(board.GP9, board.GP8)
```

2. The next step is to acquire a lock on the I2C bus. This enables us to control it:

```
while not i2c.try_lock():
    pass
```

3. Then, we make use of the scan() method to detect and print all the devices found on the I2C bus:

```
print("I2C addresses found:", [hex(device_address)
              for device_address in i2c.scan()])
```

4. Putting it all together, we have the following:

```
import time
import board
import busio

i2c = busio.I2C(board.GP9, board.GP8)

while not i2c.try_lock():
    pass

try:
    while True:
        print("I2C addresses found:", [hex(device_
address)
                for device_address in i2c.scan()])
        time.sleep(2)

finally:
    i2c.unlock()
```

Save the preceding code sample as code.py on your Pico and it should print the devices detected on the I2C bus. In the following figure, the code was used to detect the temperature sensor connected to the Pico.

```
CircuitPython REPL
I2C addresses found: ['0x38']
I2C addresses found: ['0x38']
I2C addresses found: ['0x38']
I2C addresses found: ['0x38']
I2C addresses found: ['0x38']
I2C addresses found: ['0x38']
I2C addresses found: ['0x38']
```

Figure 2.27 – Devices detected on the I2C bus

This can come in handy while troubleshooting breadboard connections. In the next section, we are going to display the temperature and humidity data using an OLED display.

Displaying temperature data using the SPI

In this section, we will make use of an OLED display that comes with an **SSD1306** driver. The **Serial Peripheral Interface (SPI)** will display the temperature and humidity data we retrieved in the previous section. Before we get started, let's take a quick look at the SPI bus.

The Serial Peripheral Interface (SPI)

The SPI is commonly used to interface peripherals such as microSD cards, memory devices, sensors, and so on. It is a synchronous interface like the I2C interface where we use a clock signal to keep the host and the peripheral in sync. The SPI bus typically consists of the following pins:

- **Chip Select (CS)** – Enables the host to select the peripheral device with which it wants to initiate communication. The chip select pin enables multiple devices on the same bus.

- **Data Out (DO)** – The pin used by the host device to transmit data to the peripheral.

- **Data In (DI)** – The pin used by the host device to receive data from the peripheral.

For a deeper dive into the SPI bus, we recommend reading through this tutorial to gain a better understanding of this interface: https://bit.ly/3fgwQud.

Wiring up the display

In this example, we are using a *128 x 32* SPI OLED Display from Adafruit (link: https://bit.ly/3yAwHJY). The display needs to be connected as shown here, where the left side of the arrow refers to the pin on the Pico while the right side refers to the OLED display's pinout:

- 3.3V → VIN

- GP2 → CLK

- GP3 → DATA

- GP11 → CS

- GP13 → D/C

- GP12 → RST

- GND

A **Fritzing** schematic of the OLED interface to the Pico is shown in the following figure:

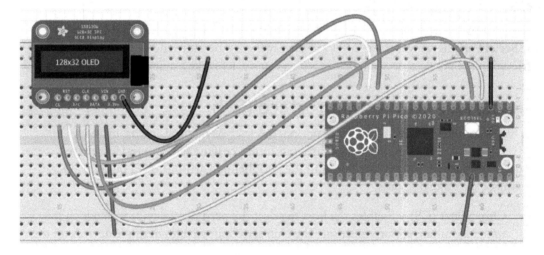

Figure 2.28 – Fritzing schematic for OLED display

Temperature Sensor Connections

In the previous schematic, we did not show the temperature sensor connections. We are assuming that you haven't disconnected the sensor.

Let's review the code needed to display the temperature and humidity data in the next section.

Displaying the temperature and humidity data

The code sample discussed in this section is available for download from here: `https://bit.ly/3byo6NA`:

1. The first step is to import the requisite modules:

    ```
    import time
    import board
    import busio
    import displayio
    import terminalio
    from adafruit_display_text import label
    import adafruit_displayio_ssd1306
    from adafruit_htu21d import HTU21D
    ```

2. The next step is to initialize the SPI bus, OLED display, I2C bus, and the temperature sensor:

    ```
    displayio.release_displays()

    spi = busio.SPI(board.GP2, MOSI=board.GP3)
    i2c = busio.I2C(board.GP9, board.GP8)

    oled_cs = board.GP11
    oled_dc = board.GP13
    oled_reset = board.GP12

    display_bus = displayio.FourWire(spi, command=oled_dc,
    chip_select=oled_cs,
                                     reset=oled_reset,
    baudrate=1000000)

    WIDTH = 128
    HEIGHT = 32

    sensor = HTU21D(i2c)
    display = adafruit_displayio_ssd1306.SSD1306(display_bus,
    width=WIDTH, height=HEIGHT)
    ```

3. Now, we can start displaying the temperature in a loop on the display:

```
while True:
    splash = displayio.Group()

    text = "Temperature: {:.1f} C".format(sensor.
temperature)
    text_area = label.Label(terminalio.FONT, text=text,
color=0xFFFFFF, x=0, y=4)
    splash.append(text_area)
    display.show(splash)

    text = "Humidity: {:.1f} %".format(sensor.relative_
humidity)
    text_area = label.Label(terminalio.FONT, text=text,
color=0xFFFFFF, x=0, y=17)
    splash.append(text_area)
    display.show(splash)

    time.sleep(2)
```

When you save the preceding code sample as code.py on your Pico, you should start seeing the temperature being continuously updated on the OLED display. Now, we will review a variant of the Pico board that comes with an onboard display.

The LILYGO RP2040 board

Since we discussed interfacing a display to the Pico via the SPI interface in the previous section, we wanted to review an alternative to the Pico board. The board (shown in the following figure) comes with a 1.14" LCD display and 4 MB of external flash.

Figure 2.29 – LILYGO RP2040 board

The board is available at `https://bit.ly/3EZvAWz` and it costs USD 11.06. You should be able to get the display on this board working with the code sample provided earlier. You just need to substitute the pin numbers (pinout shown in the following figure).

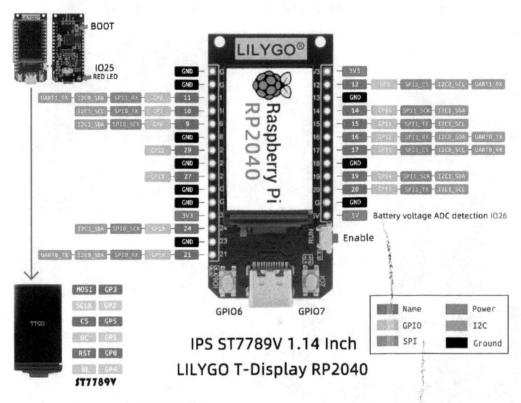

Figure 2.30 – LILYGO RP2040 pinout (image source: https://bit.ly/3EZvAWz)

In the next section, we will discuss setting up the wireless pack, which also uses the SPI. This wireless pack is going to be used across all projects that require internet connectivity in the upcoming chapters!

Setting up the wireless pack

In this section, we will discuss the setup and testing of the wireless pack. We will be using the same wireless pack for internet connectivity across all projects.

For the sake of convenience, we will make use of **Pico Omnibus – Dual Expander** from Pimoroni. The Pico and the wireless pack are mounted onto the expander as shown in the following figure:

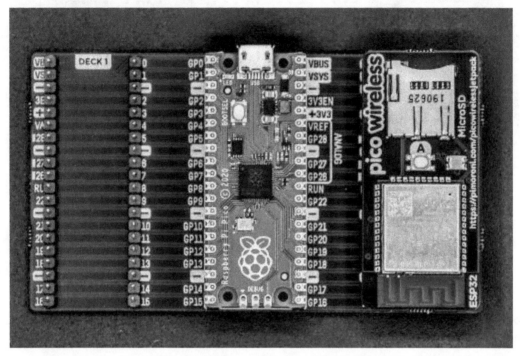

Figure 2.31 – Pico and wireless pack mounted on to a Pico Omnibus – Dual Expander

If you don't have an expander, the following pins are used to communicate with the ESP32:

- CS → GP7
- Ready → GP10
- Reset → GP11
- Clock → GP18
- MOSI → GP19
- MISO → GP16

A Fritzing schematic of the interface is shown in *Figure 2.32*:

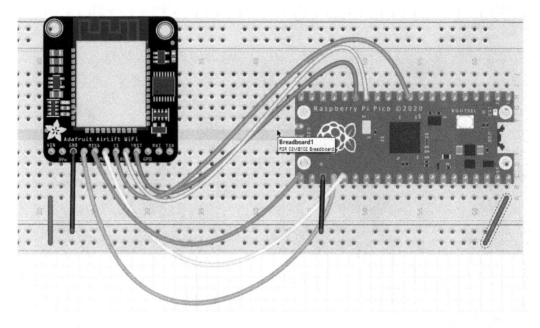

Figure 2.32 – Interfacing an ESP32 co-processor with the Raspberry Pi Pico

To connect to a Wi-Fi network using the wireless pack, create a file called `secrets.py` on your Pico and store the following information:

```
secrets = {
    'ssid' : 'home ssid',
    'password' : 'my password',
    'timezone' : "America/New_York",
    }
```

Ensure that you modify the code snippet to include your Wi-Fi credentials. It is time to test whether we can connect to the Wi-Fi network and retrieve information from a website. We have made modifications to the code from the Adafruit Learning System to test the wireless pack. The code sample is available with the chapter's downloads as `code_wifi.py` (link: `https://bit.ly/3v4Co1j`).

We have made the following changes to the code snippet:

```
esp32_cs = DigitalInOut(board.GP7)
esp32_ready = DigitalInOut(board.GP10)
esp32_reset = DigitalInOut(board.GP11)


spi = busio.SPI(board.GP18, board.GP19, board.GP16)
```

When you save the file as code.py and run it, you should see the following output:

```
CircuitPython REPL
IP lookup adafruit.com: 104.20.38.240
Ping google.com: 40 ms
Fetching text from http://wifitest.adafruit.com/testwifi/index.html
----------------------------------------
This is a test of Adafruit WiFi!
If you can read this, its working :)


----------------------------------------
```

Figure 2.33 – Wi-Fi test

Now that we have tested the wireless pack, we will move on to interfacing the NeoPixel in the next chapter.

Summary

In this chapter, we discussed interfacing two Pico boards via the UART interface and interfacing sensors via the I2C interface because they will come in handy in the projects that we are going to discuss in the upcoming chapters. The examples discussed in this chapter should help you to interface peripherals for your projects since we discussed the most used communication interfaces.

In the next chapter, we are going to discuss home automation and using the Pico in controlling appliances. You will notice that the complexity of projects increases with every chapter.

3
Home Automation Projects

In this chapter, we are going to discuss simple **home automation projects** using the Raspberry Pi Pico. We are going to work on simple projects that could be executed over a weekend and improve our immediate surroundings. We will get started by learning to interface sensors to detect events at home. Then, we will start publishing the events detected by the sensor to the cloud. We will also discuss controlling appliances at home using the Pico.

While we will be discussing the projects using the Raspberry Pi Pico, we will also introduce the **RP2040 Connect** from Arduino. The examples discussed in this chapter could be executed using an RP2040 Connect with some modifications to the code.

We are going to cover the following main topics in this chapter:

- Installing the requisite libraries
- Interfacing sensors
- Controlling appliances
- Publishing sensor events to the cloud
- Controlling LED strips
- Introducing the RP2040 Connect

Technical requirements

The following components are required for this chapter:

- Raspberry Pi Pico (link: `https://bit.ly/3AJtoAf`) – USD 4.00
- Pico Omnibus – Dual Expander (link: `https://bit.ly/3sr2GJR`) – USD 10.50
- Pico wireless pack (link: `https://bit.ly/3yPuoT9`) – USD 16.75
- IoT power relay (link: `https://bit.ly/3p9oOqo`) – USD 29.95
- Reed switch (link: `https://bit.ly/3uS3dE9`) – USD 3.95
- NeoPixel LED strip (Link: `https://bit.ly/3pp95nk`) – USD 21.95

The code samples discussed in this chapter are available from here: `https://github.com/PacktPublishing/Raspberry-Pi-Pico-DIY-Workshop/tree/main/chapter_03`.

Code in Action videos for this chapter can be viewed at `https://bit.ly/3kLsX2B`.

> **CircuitPython Installation**
>
> We are assuming that you have installed CircuitPython on your Pico. If you are not familiar with the installation process, we recommend following the installation process from *Chapter 1, Getting Started with the Raspberry Pi Pico.*

Installing the requisite libraries

In this section, we will install the requisite libraries needed for this chapter, including the NeoPixel LED and the wireless pack to the Raspberry Pi Pico. The libraries are all part of the Adafruit CircuitPython bundle. The latest bundle can be downloaded as a ZIP file from `https://circuitpython.org/libraries`. We have used the bundle version intended for CircuitPython 6.x.x.

After downloading the ZIP file, extract their contents so that we can copy the libraries we need for the project.

NeoPixel

We need the `neopixel.mpy` binary to control the NeoPixel LED. Copy over the binary to the `lib` folder.

Wireless pack

We need the `adafruit_esp32spi` library for the wireless pack. Copy over the folder (with the same name) to the `lib` folder of your Pico. We will also need the `adafruit_requests.mpy` binary from the bundle.

> **Setting Up the Wireless Pack**
>
> In *Chapter 2, Serial Interfaces and Applications*, we discussed setting up the wireless pack in detail. We recommend that you read through this chapter and then set up your wireless pack.

Interfacing sensors

In this section, we are going to discuss **interfacing sensors** on the Pico, in particular, a **reed switch**. We picked a reed switch because we have all been in situations where we either cannot hear a door open, or you are not aware when someone doesn't close a door properly. We are going to discuss detecting such situations using a reed switch.

A reed switch is a type of switch that closes and makes contact in the presence of a magnetic field. *Figure 3.1* shows a reed switch sold by Adafruit that could be used as a door sensor. The sensor consists of two pieces, namely, the magnet and the reed switch.

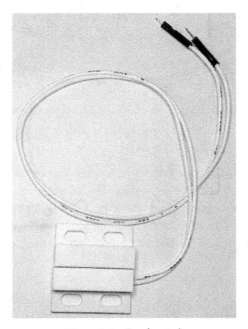

Figure 3.1 – Reed switch

The magnet is usually installed on the door while the reed switch is installed on the door frame. When the magnet is within 13 mm of the reed switch, the switch is closed. We are going to make use of this principle to detect the door state.

The steps to interface a door sensor with the Pico include the following:

1. The first step is to interface the door sensor with the Pico, as shown in the following diagram:

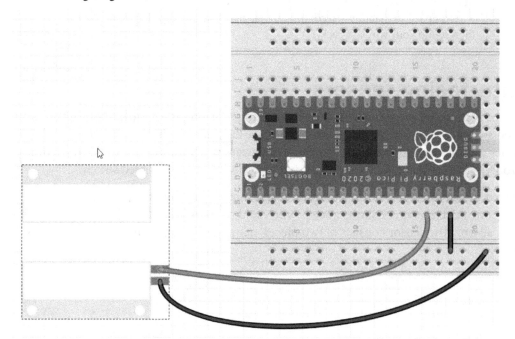

Figure 3.2 – Fritzing schematic for connecting the door sensor to the Pico

In the schematic, one wire of the reed switch is connected to the *GP12 pin*, while the other end is connected to the *GND* pin.

2. We are going to discuss the code sample with the Python interpreter. The code sample discussed in this section is available for download along with this chapter as `code_door_sensor.py`. We recommend launching the Python interpreter on your Pico and working along with this example. Open the serial terminal on the Mu **Integrated Development Environment** (**IDE**) from the toolbar on the top of the IDE.

Figure 3.3 – Location of the Serial button on the toolbar

3. If you see the output as shown in the following screenshot, press any key to enter the Python interpreter:

```
CircuitPython REPL
Auto-reload is on. Simply save files over USB to run them or enter REPL to disable.

Press any key to enter the REPL. Use CTRL-D to reload.
|
```

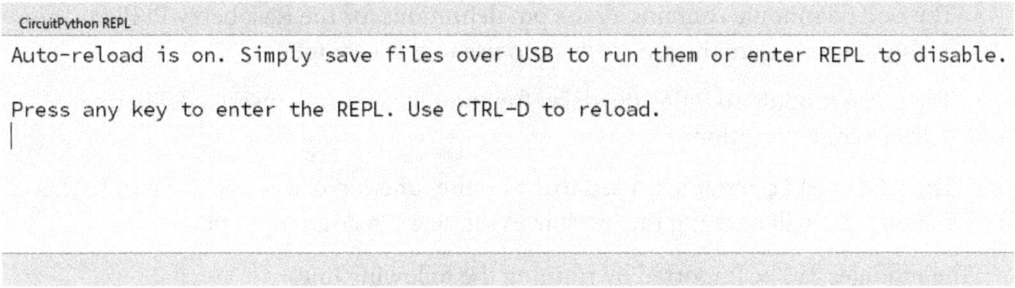

Figure 3.4 – Serial terminal on the Mu IDE

4. If there is code running on the Pico, there might be a blank screen or something being printed on the serial terminal (as shown in the following screenshot). Press *Ctrl + C* to interrupt code execution.

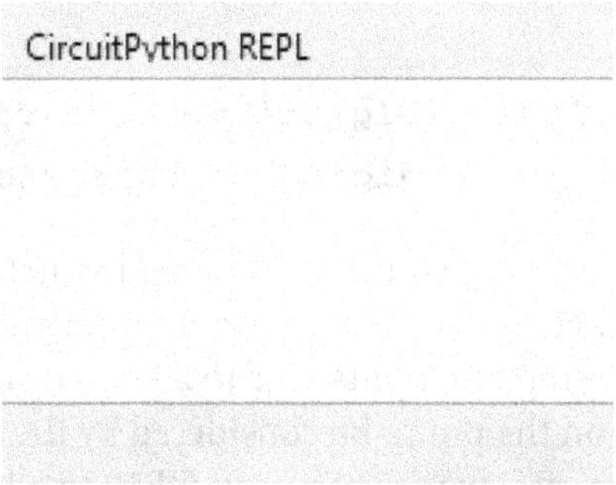

```
CircuitPython REPL
```

Figure 3.5 – Serial terminal with no output on the Mu IDE

5. The Python interpreter would look like something similar to that shown in the following screenshot:

```
CircuitPython REPL
Auto-reload is on. Simply save files over USB to run them or enter REPL to disable.

Press any key to enter the REPL. Use CTRL-D to reload.

Adafruit CircuitPython 6.2.0-beta.4 on 2021-03-18; Raspberry Pi Pico with rp2040
>>>
```

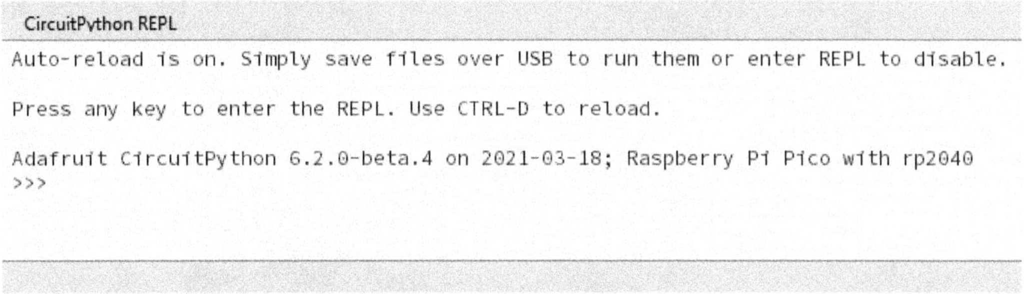

Figure 3.6 – Python interpreter on the Pico

6. The first step is to import the requisite modules. For this example, we will import the `board`, `time`, and `digitalio` modules:

    - The `board` module contains all the pin definitions for the Raspberry Pi Pico. This enables us to declare the pin we want to use in this example.

    - The `time` module contains the delay function. It is used to insert a deliberate delay in code execution.

    - The `digitalio` module is used to set up the function of a particular pin. In this example, we will be using the module to set up a pin as an input pin.

    The modules can be imported by running the following lines:

    ```
    >>> import board
    >>> import digitalio
    >>> import time
    ```

7. Since the reed switch is connected to the *GP12* pin, we are going to initialize it as an input pin:

    ```
    >>> switch = digitalio.DigitalInOut(board.GP12)
    >>> switch.direction = digitalio.Direction.INPUT
    ```

8. The *GP12* pin (to which we have connected the reed switch) is floating, and we need to tie it to a known state. Hence, we will use an internal pull-up resistor – a resistor. Pull-up resistors enable tying the inputs pins to a known state. This enables the prevention of any noise on the pin to be considered by the code as a change in the status of the door. In this case, we are tying pin GP12 to a high state:

    ```
    >>> switch.pull = digitalio.Pull.UP
    ```

    The following diagram shows the GP12 pin in a pull-up configuration:

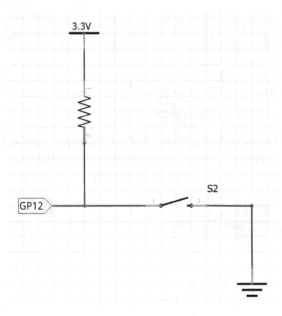

Figure 3.7 – GPIO pin pull-up configuration

9. Alternatively, you could tie the GPIO pin to a pull-down state, and the GPIO pin configuration is set as follows:

```
>>> switch.pull = digitalio.Pull.DOWN
```

The following diagram shows the GP12 pin in a pull-down configuration:

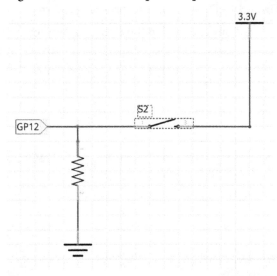

Figure 3.8 – GPIO pin pull-down configuration

10. We are going to set *GP12* in the pull-up state. Now, let's read the switch state in an infinite loop and test it with the door magnet as follows:

```
>>> while True:
...         print(switch.value)
...         time.sleep(1)
```

11. Now, the terminal should start printing the door sensor's state at a *1-second* interval. When the magnet is present, in other words, the door is closed, the program prints `False` to the screen. When the door is open, it prints `True` to the screen, as can be seen on the following screen:

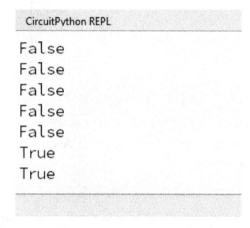

Figure 3.9 – Door sensor states printed to the screen

12. Now that we have tested our code using the Python interpreter, we can save the code as `code.py` and observe the door sensor's state printed in an infinite loop:

```python
import board
import digitalio
import time


switch = digitalio.DigitalInOut(board.GP12)
switch.direction = digitalio.Direction.INPUT
switch.pull = digitalio.Pull.UP


while True:
    print(switch.value)
    time.sleep(1)
```

Have you ever wondered where such door sensors are used? Have you heard door chimes while walking into a convenience store? They use some form of door sensor. In the following photo, you can see a door sensor being used to monitor an access-controlled gate:

Figure 3.10 – Door sensor by an access-controlled gate

We are going to make use of this binary output from the door sensor to detect our door status and report events to the cloud or control appliances. In the next section, we are going to discuss controlling appliances inside a room based on the door sensor's state.

Controlling appliances

In this section, we are going to **control appliances** using a product called **IoT Power Relay** (shown in the following image). We are going to make some minor tweaks to the code sample discussed in the previous section to turn on a light when the door is open and vice-versa.

> **Appliances Controlled by the Power Relay**
>
> We recommend using the power relay for your projects because it offers a safe way to control AC appliances. We recommend controlling simple resistive loads such as a floor lamp.

The following photo shows the IoT power relay. The terminal on the bottom side of the picture is used to control the relay.

Figure 3.11 – IoT power relay (image source: sparkfun.com. License: CC by 2.0)

The IoT power relay enables controlling **alternating current** (**AC**) appliances with a *3.3 V* signal. It comes with the requisite protection to safely interface the Pico with the relay. We will be making use of the *GP11* pin to control the relay. The schematic to control the power relay is shown in the following diagram. We are going to connect the positive terminal to *GP11* and the negative to *GND*.

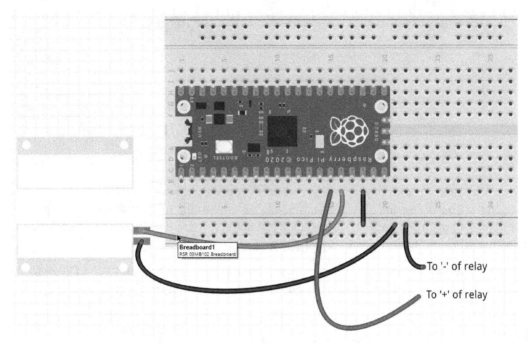

To '-' of relay

To '+' of relay

Figure 3.12 – Fritzing schematic to interface the power relay

We are going to make minor changes to the code sample discussed in the previous section to control an appliance. Let's do the following:

1. The first step is to initialize pin *GP11* as an OUTPUT pin:

```
>>> relay = digitalio.DigitalInOut(board.GP11)
>>> relay.direction = digitalio.Direction.OUTPUT
```

2. Now, let's modify the `while` loop to turn on the relay based on door state. In the previous section, we noticed that the door is open if the switch state is True. We need to turn on the lights if the switch state is True:

```
import board
import digitalio
import time

switch = digitalio.DigitalInOut(board.GP12)
switch.direction = digitalio.Direction.INPUT
switch.pull = digitalio.Pull.UP

relay = digitalio.DigitalInOut(board.GP11)
```

```
relay.direction = digitalio.Direction.OUTPUT

while True:
    if switch.value:
        relay.value = True
    else:
        relay.value = False
    time.sleep(1)
```

When you power up the relay using a power cord, you should hear the relay being energized with a *tick* sound. Try connecting a table lamp to the relay. It should light up whenever you open the door. In case you are not familiar with relays, we recommend the following article: https://bit.ly/3qQRzv9.

We have reviewed controlling appliances based upon sensor input to the RP2040 Pico. In the next section, we are going to discuss publishing the door sensor states to the cloud.

Publishing sensor events to the cloud

In this section, we are going to discuss **publishing sensor events** to the cloud. This is especially important to receive email alerts and mobile notifications when a door is open. We are assuming that you have set up your wireless pack following the instructions from *Chapter 2, Serial Interfaces and Applications*.

Setting up Adafruit IO

In this section, we will discuss setting up Adafruit IO for our project. Adafruit IO is a cloud service provided by Adafruit Industries that enables the Raspberry Pi Pico to be connected to the internet and sensor events published, as well as appliances controlled from anywhere on the web. The setup steps include the following:

Additional Resource

We wrote this section using https://learn.adafruit.com/ as a resource. If you run into problems, refer to the links provided in each step.

---

1. The first step is to create an account for yourself at https://io.adafruit.com/. While the basic plan is free, there is also a subscription available as follows:

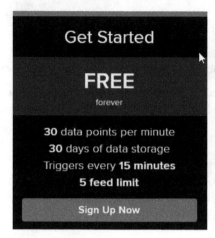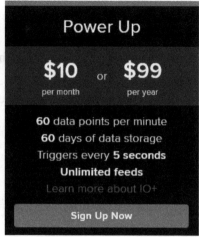

Figure 3.13 – Signing up for adafruit.com

2. Once you have created an account, retrieve your key from **My Key**, as shown in *Figure 3.14*. Copy the username and the active key information (link: `https://bit.ly/3vK6QfK`).

Figure 3.14 – Retrieving the Adafruit IO key

3. We need to save the retrieved information to `secrets.py` as follows (link: `https://bit.ly/2SFbkHn`):

```
secrets = {
    'ssid' : '_your_wifi_ssid',
    'password' : '_your_wifi_password',
    'aio_username' : '_your_adafruit_io_username',
    'aio_key' : '_your_big_huge_super_long_aio_key_'
    }
```

In the next section, we are going to discuss publishing events to the Adafruit IO cloud service.

Publishing events

We are now going to review publishing events using the Raspberry Pi Pico. The steps include the following:

1. Download the code bundle from `https://bit.ly/3gNPadR` and extract the files.

2. Copy the contents of the `lib` folder to the `lib` folder of the CircuitPython device.

3. We are going to modify the script discussed in the previous section to publish a random value to the *temperature feed*. The first step is to import the requisite modules:

```
import board
import busio
from digitalio import DigitalInOut
import adafruit_requests as requests
import adafruit_esp32spi.adafruit_esp32spi_socket as
socket
from adafruit_esp32spi import adafruit_esp32spi
from adafruit_io.adafruit_io import IO_HTTP, AdafruitIO_
RequestError
```

4. Then, we import the Wi-Fi credentials for the project as follows:

```
# Get wifi details and more from a secrets.py file
try:
    from secrets import secrets
except ImportError:
    print("WiFi secrets are kept in secrets.py, please
add them there!")
    raise
```

5. Next, we initialize the Wi-Fi module:

```
#   ESP32 pins
esp32_cs = DigitalInOut(board.GP7)
esp32_ready = DigitalInOut(board.GP10)
esp32_reset = DigitalInOut(board.GP11)


spi = busio.SPI(board.GP18, board.GP19, board.GP16)
```

```
esp = adafruit_esp32spi.ESP_SPIcontrol(spi, esp32_cs,
esp32_ready, esp32_reset)
```

```
requests.set_socket(socket, esp)
```

```
if esp.status == adafruit_esp32spi.WL_IDLE_STATUS:
    print("ESP32 found and in idle mode")
print("Firmware vers.", esp.firmware_version)
print("MAC addr:", [hex(i) for i in esp.MAC_address])
```

```
for ap in esp.scan_networks():
    print("\t%s\t\tRSSI: %d" % (str(ap['ssid'], 'utf-8'),
ap['rssi']))
```

6. Then, we connect to the network based on the information retrieved from
    `secrets.py`:

```
print("Connecting to AP...")
while not esp.is_connected:
    try:
        esp.connect_AP(secrets["ssid"],
secrets["password"])
    except RuntimeError as e:
        print("could not connect to AP, retrying: ", e)
        continue
print("Connected to", str(esp.ssid, "utf-8"), "\tRSSI:",
esp.rssi)
print("My IP address is", esp.pretty_ip(esp.ip_address))
```

7. We initialize a socket connection and initialize an *Adafruit IO object* as follows:

```
socket.set_interface(esp)
requests.set_socket(socket, esp)
```

```
aio_username = secrets["aio_username"]
aio_key = secrets["aio_key"]
```

```
# Initialize an Adafruit IO HTTP API object
io = IO_HTTP(aio_username, aio_key, requests)
```

8. We then create a new temperature feed if one doesn't exist:

```
try:
    # Get the 'temperature' feed from Adafruit IO
    temperature_feed = io.get_feed("temperature")
except AdafruitIO_RequestError:
    # If no 'temperature' feed exists, create one
    temperature_feed = io.create_new_feed("temperature")
```

9. We publish a random value to the temperature feed, as demonstrated here:

```
# Send random integer values to the feed
random_value = randint(0, 50)
print("Sending {0} to temperature feed...".format(random_
value))
io.send_data(temperature_feed["key"], random_value)
print("Data sent!")
```

10. We retrieve the published value to test whether everything works:

```
# Retrieve data value from the feed
print("Retrieving data from temperature feed...")
received_data = io.receive_data(temperature_feed["key"])
print("Data from temperature feed: ", received_
data["value"])
```

You can go to https://io.adafruit.com/ to view the published feed. In the next section, we will discuss the next steps for publishing events.

Next steps

In the previous section, we discussed connecting the Pico to a Wi-Fi network and publishing random feeds. You can use the same script to publish door sensor events, remotely monitor plants, and so on. Check out this book's repository for some examples.

Controlling LED strips

In this section, we will interface a **NeoPixel LED strip** with the Pico. NeoPixels are serially connected LEDs that are individually addressable. The NeoPixels come in a variety of form factors, namely, horizontal bars, flexible strips, and circular rings. We will be discussing this example with a **NeoPixel ring** (link: `https://bit.ly/3cn5pxj`). We are discussing the LED strip interface because they could make a great holiday lighting project or an ambient light controller. Our favorite project using Pico and the NeoPixel LED is this table lamp (link: `https://bit.ly/3r6iIdJ`).

The examples discussed in this section make use of helper functions from Adafruit.

The NeoPixel ring requires three connections, namely, **Data IN or DIN**, **Power**, and **Ground** (connections shown in *Figure 3.23*). While the Power pin is connected to the *3.3V* pin of the Pico, the Data IN pin is connected to *GP10* of the Pico.

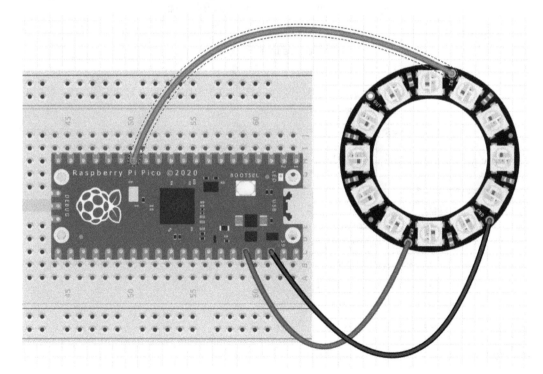

Figure 3.15 – Schematic for connecting the Pico to the NeoPixel ring

To control the NeoPixel LED ring using the Pico, we need to do the following:

1. The first step with the code is to import the requisite modules as follows:

```
import time
import board
import neopixel
```

2. Then, we initialize the NeoPixel object and set *GP10* as the control pin:

```
pixel_pin = board.GP10
num_pixels = 12
pixels = neopixel.NeoPixel(pixel_pin, num_pixels,
brightness=0.3, auto_write=False)
```

Let's make use of a helper function from Adafruit as follows:

```
def color_chase(color, wait):
    for i in range(num_pixels):
        pixels[i] = color
        time.sleep(wait)
        pixels.show()
    time.sleep(0.5)
```

3. Now, we will enter the `while` loop while we switch between colors as follows:

```
while True:
    color_chase(RED, 0.1)   # Increase the number to slow
down the color chase
    color_chase(YELLOW, 0.1)
    color_chase(GREEN, 0.1)
    color_chase(CYAN, 0.1)
    color_chase(BLUE, 0.1)
    color_chase(PURPLE, 0.1)
```

4. Putting it all together, we have the following:

```
import time
import board
import neopixel

pixel_pin = board.GP10
```

```
num_pixels = 12

pixels = neopixel.NeoPixel(pixel_pin, num_pixels,
brightness=0.3, auto_write=False)

def color_chase(color, wait):
    for i in range(num_pixels):
        pixels[i] = color
        time.sleep(wait)
        pixels.show()
    time.sleep(0.5)

RED = (255, 0, 0)
YELLOW = (255, 150, 0)
GREEN = (0, 255, 0)
CYAN = (0, 255, 255)
BLUE = (0, 0, 255)
PURPLE = (180, 0, 255)

while True:
    color_chase(RED, 0.1)  # Increase the number to slow
down the color chase
    color_chase(YELLOW, 0.1)
    color_chase(GREEN, 0.1)
    color_chase(CYAN, 0.1)
    color_chase(BLUE, 0.1)
    color_chase(PURPLE, 0.1)
```

In this example, we are making use of a NeoPixel ring with 12 LEDs. You will have to modify the LED count before saving the code sample on the Pico. You can daisy chain multiple LED strips together and install them along the corners of the wall in your home. You can set the ambient lighting to different colors based on the time of day.

In the next section, we will introduce the RP2040 Connect, a variant of the Raspberry Pi Pico.

Introducing the RP2040 Connect

In this section, we will introduce the **RP2040 Connect** from the Arduino foundation. The RP2040 Connect is a variant of the Pico in the Arduino Nano form factor. We like this board because it comes with Bluetooth and Wi-Fi connectivity. It also comes with an onboard accelerometer, gyroscope, RGB LED, and a microphone. This provides tremendous opportunities for prototyping with the RP2040 Connect. The following diagram shows the pinouts for the RP2040 Connect (available for purchase from here: https://bit.ly/2YZClrY):

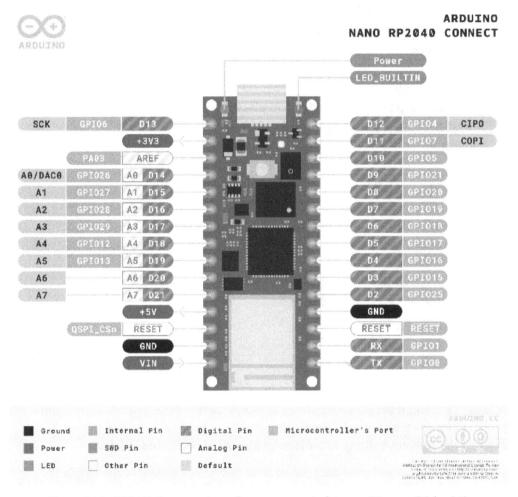

Figure 3.16 – RP2040 Connect pinout (image source: Arduino.cc. License: CC-by-SA)

In the next section, we will discuss installing CircuitPython on the RP2040 Connect.

Installing CircuitPython on the RP2040 Connect

Installing CircuitPython is the same as what we discussed in *Chapter 1*, *Getting Started with the Raspberry Pi Pico*, except that there will be an extra step in these instructions where we enumerate the flash drive. The steps to install CircuitPython on the RP2040 Connect are as follows:

1. Set up your RP2040 Connect on a breadboard, as shown in the following photo:

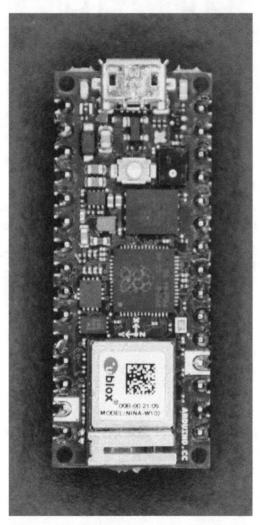

Figure 3.17 – RP2040 Connect on a breadboard

2. Tap on the reset button twice to put the device into bootloader mode (highlighted using the rectangle).

Figure 3.18 – Reset button on the RP2040 Connect

3. The device enumerates as a storage device. Download the CircuitPython binary from `https://bit.ly/34R29pJ` and copy it over to the storage device.

4. Upon flashing, the device resets itself and enumerates as a storage device with the name of `CIRCUITPY`.

Figure 3.19 – RP2040 Connect flashed with the CircuitPython binary

5. We should be able to program the RP2040 Connect using the Mu IDE. In the following snapshot, we can see that the IDE has detected a CircuitPython device:

Detected new CircuitPython device: Adafruit CircuitPlayground.

Figure 3.20 – RP2040 Connect detected by the Mu IDE

6. We can open a serial connection to start programming in CircuitPython. In the following screenshot, you will notice the revision of CircuitPython running on the Arduino RP2040 Connect:

```
CircuitPython REPL
Auto-reload is on. Simply save files over USB to run them or enter REPL to disable.

Press any key to enter the REPL. Use CTRL-D to reload.

Adafruit CircuitPython 6.3.0-rc.0 on 2021-05-25; Arduino Nano RP2040 Connect with rp2040
>>>
```

Figure 3.21 – Python interpreter running on the RP2040 Connect

As you might have observed, the installation process for the RP2040 Connect is very similar to that of the Raspberry Pi Pico. Let's discuss how to connect it to the internet using the onboard Wi-Fi module to demonstrate the ease of programming in CircuitPython using the RP2040 Connect.

Connecting the RP2040 to the internet

In this section, we are going to discuss connecting the RP2040 Connect to the internet. We are basically going to make some minor modifications to the code sample (`code_wifi.py` – available with this chapter's downloads) discussed in the previous chapter. The modified code sample is available for download as `code_wifi_rp2040_connect.py`.

Installing prerequisites

The prerequisites for setting up the onboard Wi-Fi module are the same as that of the ESP32 wireless pack used with the Pico. From the CircuitPython library bundle downloaded earlier, we need `adafruit_esp32spi`. Copy over the folder (with the same name) to the `lib` folder of your Pico. We will also need the `adafruit_requests.mpy` binary from the bundle. We also need to save the Wi-Fi credentials in a file called `secrets.py` in the following format:

```
secrets = {
    'ssid' : 'SSID',
    'password' : 'Password',
    'timezone' : "America/New_York",
    }
```

Do not forget to include your Wi-Fi credentials in the file. Now, we will modify the code sample, `code_wifi.py`.

Modifying the code

The modifications required to `code_wifi.py` are minimal. We will just be modifying the pin definitions between lines 25 and 29. The original definition was as follows:

```
esp32_cs = DigitalInOut(board.GP7)
esp32_ready = DigitalInOut(board.GP10)
esp32_reset = DigitalInOut(board.GP11)


spi = busio.SPI(board.GP18, board.GP19, board.GP16)
```

It is modified to the Arduino RP2040 Connect's pin definitions:

```
#  ESP32 pins
esp32_cs = DigitalInOut(board.CS1)
esp32_ready = DigitalInOut(board.ESP_BUSY)
esp32_reset = DigitalInOut(board.ESP_RESET)


spi = busio.SPI(board.SCK1, board.MOSI1, board.MISO1)
```

When you save the file as `code.py` and run it, you should see the following output:

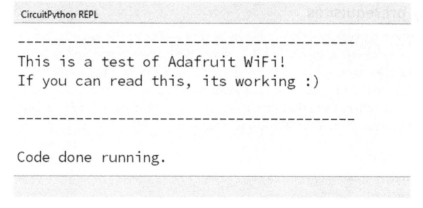

Figure 3.22 – RP2040 Connect Wi-Fi test

From this example, you might have noticed that programming in CircuitPython is consistent across any variant of the Raspberry Pico, except for some minor modifications to the code.

> **Challenge to the Reader**
>
> If you happen to own the RP2040 Connect, pick any code sample discussed earlier in the chapter and try using it with the RP2040 Connect. Make sure that you have the board's pinouts to make modifications to your code.

Summary

In this chapter, we discussed interfacing sensors to the **General Purpose Input/Output (GPIO)** pins of the Pico, turning appliances on/off using a relay box, interfacing LED strips to the Pico, and publishing sensor events to the cloud. We discussed controlling an LED strip using a variant of the Pico, namely, the RP2040 Connect.

The examples discussed in this chapter were primarily chosen as examples that could be discussed as individual weekend projects.

In the next chapter, we will have fun with gardening. See you shortly!

4
Fun with Gardening!

In this chapter, we will discuss interfacing a soil sensor to a Raspberry Pi Pico and publishing the soil moisture and temperature values to ThingSpeak (`https://thingspeak.com`), an **Internet of Things (IoT)** platform that enables you to visualize collected data. By the end of this chapter, you should be able to build something like this:

Figure 4.1 – Fun with gardening

In the preceding photo, a STEMMA soil sensor from Adafruit Industries is inserted next to a small potted plant. The soil sensor is interfaced to a Raspberry Pi Pico. The Pico is mounted onto a Dual Expander pack. A Pico wireless pack is interfaced to the left of the Pico for uploading soil moisture values to ThingSpeak.

We are going to discuss the following topics in this chapter:

- Why gardening?
- Installing the requisite libraries
- Setting up the soil sensor
- Setting up the wireless pack
- Setting up the NeoPixel **light-emitting diode** (**LED**)
- Publishing data to ThingSpeak
- Putting it all together

Let's get started!

Technical requirements

The following hardware is recommended for this chapter:

- Raspberry Pi Pico (`https://bit.ly/3AJtoAf`)—**United States Dollars** (**USD**) 4
- Adafruit STEMMA soil sensor (`https://www.adafruit.com/product/4026`)—USD 7.50
- Pico Omnibus—Dual Expander (`https://bit.ly/3sr2GJR`)—USD 10.50
- Pico wireless pack (`https://bit.ly/3yPuoT9`)—USD 16.75
- **Inter-Integrated Circuit** (**I2C**) STEMMA cable (`https://bit.ly/3shASra`)—USD 1.50
- NeoPixel LED (optional) (`https://bit.ly/3ALSazI`)—USD 5.50

You will also need a potted plant of your choice. The code samples for this chapter are available at the following link: `https://github.com/PacktPublishing/Raspberry-Pi-Pico-DIY-Workshop/tree/main/chapter_04`.

Code in Action videos for this chapter can be viewed at `https://bit.ly/3w3CqY9`.

> **CircuitPython Installation**
>
> We are assuming that you have installed CircuitPython on your Pico. If you are not familiar with the installation process, we recommend following the installation process from *Chapter 1, Getting Started with the Raspberry Pi Pico.*

Why gardening?

Why did we choose gardening for our project in this chapter?

Gardening is one of the oldest hobbies. It has periods of upswings, as was the case with the pandemic. While gardening can be embraced at many levels, technology can enhance it, reduce fear of failure, and serve as an educational tool. Sensors and response mechanisms such as watering or lighting tools can help with plants that require extra care, including some houseplants, water plants, orchids, and bonsai trees.

Besides being a generic hobby, gardening can also be used to grow food efficiently in resource-starved environments such as urban settings. With more people living in such settings, it is increasingly important for people to be able to grow at least some of their own food. Homegrown food can also lead to healthy eating and living habits and provide other advantages such as reducing stress and improving mental health. Gardening and urban agriculture on any scale can also help endangered bees, butterflies, and other species.

Plants require soil, water, light in specific wavelengths for specific periods of time, and fertilizers. Sensors can be used to measure how much light a plant is getting at a specific location, humidity in the soil, soil **potential of hydrogen** (**pH**), and similar parameters. Sensors can also be used to measure environmental temperature, carbon dioxide, and **volatile organic compounds** (**VOCs**), all of which may have an impact on the growth of plants. Once the outputs of sensors are collected and interpreted by tools such as the Raspberry Pi Pico, plants can be provided with required inputs, such as water, fertilizer, adequate light, or absence of light—say, in the form of a motorized black curtain—and so on.

On a larger scale, the Raspberry Pi Pico can be used for urban agriculture, or any agricultural setting, including hydroponics, vertical farming, or generic farming in developing nations and economically disadvantaged communities. A network of Pico boards can serve as edge devices, sensing problems and providing nutrition or other support for the survival and growth of plants. Similarly, the work can be extended to urban gardens or forestry.

Installing the requisite libraries

In this section, we will install the requisite libraries needed for interfacing the soil sensor, NeoPixel LED, and the wireless pack to the Raspberry Pi Pico. The libraries are all part of the Adafruit CircuitPython bundle. The latest bundle can be downloaded as a ZIP file from `https://circuitpython.org/libraries`. We used the bundle version meant for CircuitPython 6.x.x.

After downloading the ZIP file, extract its contents so that we can copy the libraries we need for the project.

Soil sensor

We are using a capacitive soil sensor based on the **Adafruit seesaw** framework. You can see this in the following screenshot:

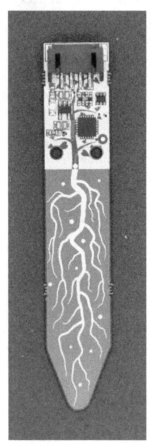

Figure 4.2 – Capacitive soil sensor from Adafruit

If you are not familiar with Adafruit seesaw, this is a framework that enables the interfacing of any type of sensor, drive motor, control **pulse-width modulation (PWM)** output, and so on via the I2C interface. This is especially helpful in scenarios where your microcontroller does not come with the required peripheral. In this project, the capacitive soil sensor is interfaced to a microcontroller running the seesaw firmware. This simplifies the problem, as we can read the soil moisture and humidity values using the `adafruit_seesaw` libraries. You can learn more about the seesaw framework here: `https://bit.ly/3D0qN7c`. Two libraries are needed for interfacing the soil sensor: `adafruit_seesaw` and `adafruit_bus_device`.

In the extracted library bundle, there are two folders—namely, `adafruit_seesaw` and `adafruit_bus_device`. Copy over the folders to the `lib` folder of your Pico.

Wireless pack

We need the `adafruit_esp32spi` library for the wireless pack. Copy over the folder (with the same name) to the `lib` folder of your Pico. We will also need the `adafruit_reqests.mpy` binary from the bundle.

NeoPixel

We need the `neopixel.mpy` binary to control the NeoPixel LED (shown in the following screenshot). Copy over the binary to the `lib` folder:

Figure 4.3 – Red, green, blue, white (RGBW) NeoPixel LED

The Pico's `lib` folder contents should look something like this:

Name	Date modified	Type	Size
adafruit_bus_device	8/8/2021 6:58 PM	File folder	
adafruit_esp32spi	8/8/2021 6:58 PM	File folder	
adafruit_seesaw	8/8/2021 5:33 AM	File folder	
adafruit_requests.mpy	8/6/2021 5:10 AM	MPY File	14 KB
neopixel.mpy	8/6/2021 5:09 AM	MPY File	3 KB

Figure 4.4 – Libraries needed for the current project

Next, we will set up the individual components for the project and test them. We will get started with the soil sensor.

Setting up the soil sensor

In this section, we will interface and test the soil sensor to the Pico. A *Fritzing* schematic of the soil sensor is shown here:

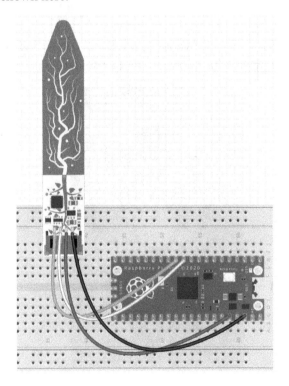

Figure 4.5 – Fritzing schematic to interface the soil sensor to the Pico

> **Pico Pinout**
>
> A pinout reference card for the Pico will come in handy while wiring up components. We provided a link to the pinout card in *Chapter 1, Getting Started with the Raspberry Pi Pico*.

The soil sensor comes with an I2C interface, and the STEMMA soil sensor connector consists of four pins—namely *SCL*, *SDA*, *VIN*, and *GND*. If you are not familiar with the I2C interface, we recommend reading *Chapter 2* of this book, *Serial Interfaces and Applications*. In the *Fritzing* schematic, looking from left to right, the soil sensor is connected to the Pico as follows:

- SCL to the GP9 pin of the Pico.
- SDA to the GP8 pin of the Pico.
- VIN to the 3.3V pin of the Pico.
- The GND pins are tied together.

> **Pull-Up Resistors for I2C Bus**
>
> Typically, pull-up resistors are needed for the SCL and SDA pins of the I2C bus. The STEMMA soil sensor comes with pull-up resistors on the I2C bus. While using off-the-shelf sensors that come with an I2C interface, read the schematic and understand the I2C bus configuration of the sensor. If you are not familiar with the need for pull-up resistors for the I2C bus, we recommend checking out *Chapter 2, Serial Interfaces and Applications*.

Now that we have wired up the soil sensor, let's write some code to test it.

> **Code Samples**
>
> The code samples used to discuss the testing of the individual components are based on the *Adafruit Learning System*, which is an excellent resource for any hardware project.

Open code.py (located on our Pico) and make the following modifications:

1. The first step is to import the requisite modules, as follows:

```
import time
import board
import busio
from adafruit_seesaw.seesaw import Seesaw
```

2. This is followed by initializing the I2C bus by specifying the SCL pin (GP9) and the SDA pin (GP8). Then, we instantiate an object belonging to the `Seesaw` class to communicate with the soil sensor. The code is illustrated in the following snippet:

```
i2c = busio.I2C(board.GP9, board.GP8)
ss = Seesaw(i2c, addr=0x36)
```

3. Now, we can read the soil moisture levels and temperature, as follows:

```
touch = ss.moisture_read()
temp = ss.get_temp()
```

4. We can read the moisture levels and temperature in an infinite loop and print this to the serial port. Putting it all together, we have the following:

```
import time
import board
import busio
from adafruit_seesaw.seesaw import Seesaw

i2c = busio.I2C(board.GP9, board.GP8)
ss = Seesaw(i2c, addr=0x36)

while True:
    # read moisture level through capacitive touch pad
    touch = ss.moisture_read()

    # read temperature from the temperature sensor
    temp = ss.get_temp()

    print("temp: " + str(temp) + "  moisture: " +
str(touch))
    time.sleep(1)
```

Save the code file as `code.py` and you should be able to see the soil sensor output, as shown in the following screenshot. Since it is a capacitive sensor, touching the sensor could cause a change in the measured capacitance. You can observe the change in capacitance levels when the sensor was touched:

```
CircuitPython REPL
temp:  27.457   moisture:  353
temp:  28.0913  moisture:  1014
temp:  29.2563  moisture:  1014
temp:  29.8905  moisture:  1014
temp:  30.1041  moisture:  1014
temp:  30.0005  moisture:  352
temp:  29.787   moisture:  352
```

Figure 4.6 – Soil sensor output

The soil moisture levels range from 200 (very dry) to 2,000 (very wet). The levels depend upon several factors, including soil type, temperature, and so on.

In the next section, we are going to set up the wireless pack.

Setting up the wireless pack

In this section, we will discuss setting up and testing the wireless pack. For the sake of convenience, we will make use of **Pico Omnibus (Dual Expander)** from Pimoroni. The Pico and the wireless pack are mounted onto the expander, as shown in the following screenshot:

Figure 4.7 – Pico and wireless pack mounted onto Pico Omnibus (Dual Expander)

> **Breadboard Connections**
>
> If you don't have an expander, the following pins are used to communicate with the ESP32: CS → GP7; Ready → GP10; Reset → GP11; Clock → GP18; **Master Out Slave In (MOSI)** → GP19; **Master In Slave Out (MISO)** → GP16. A *Fritzing* schematic of the interface is shown here:

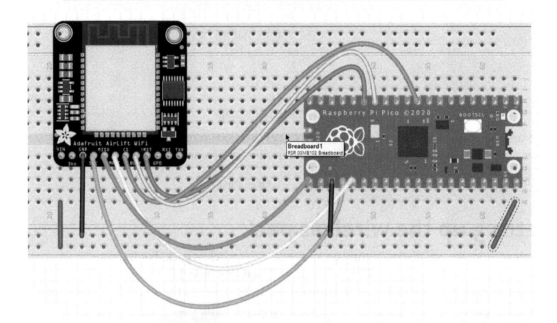

Figure 4.8 – Interfacing an ESP32 co-processor to the Raspberry Pi Pico

To connect to a Wi-Fi network using the wireless pack, create a file called `secrets.py` on your Pico and store the following information in it:

```
secrets = {
    'ssid' : 'home ssid',
    'password' : 'my password',
    'timezone' : "America/New_York",
}
```

Ensure that you modify the code snippet to include your Wi-Fi credentials. It's now time to test whether we can connect to the Wi-Fi network and retrieve information from a website. We have made modifications to the code from the *Adafruit Learning System* to test the wireless pack. The code sample is available with the chapter's downloads as `code_wifi.py` (see `https://bit.ly/3v4Co1j`).

We have made the following changes to the code:

```
esp32_cs = DigitalInOut(board.GP7)
esp32_ready = DigitalInOut(board.GP10)
esp32_reset = DigitalInOut(board.GP11)


spi = busio.SPI(board.GP18, board.GP19, board.GP16)
```

When you save the file as code.py and run it, you should see the following output:

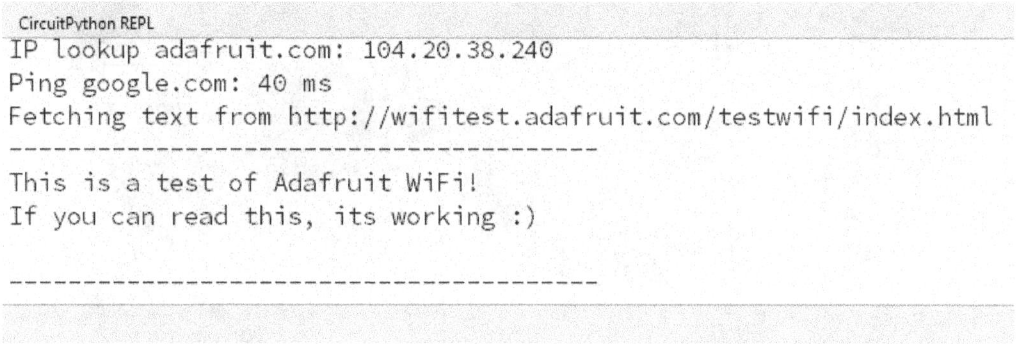

```
CircuitPython REPL
IP lookup adafruit.com: 104.20.38.240
Ping google.com: 40 ms
Fetching text from http://wifitest.adafruit.com/testwifi/index.html
----------------------------------------
This is a test of Adafruit WiFi!
If you can read this, its working :)


----------------------------------------
```

Figure 4.9 – Wi-Fi test

Now that we have tested the wireless pack, we will move on to interfacing the NeoPixel LED in the next section.

Setting up the NeoPixel LED

In this section, we will interface and test a NeoPixel LED. We are adding an LED to the project because it could be used as a visual aid to let you know if your soil is too dry. If you are not familiar with NeoPixel LEDs, these are individually addressable RGB LEDs. You can find more information about them here: https://bit.ly/3sin0x5.

The NeoPixel is connected to GP0, as shown in *Figure 4.7*. The connections between the Pico and the NeoPixel are listed as follows:

- GP0 → *In* pin of NeoPixel.
- VBUS → + pin.
- The GND pins are tied together.

An interface diagram is shown here:

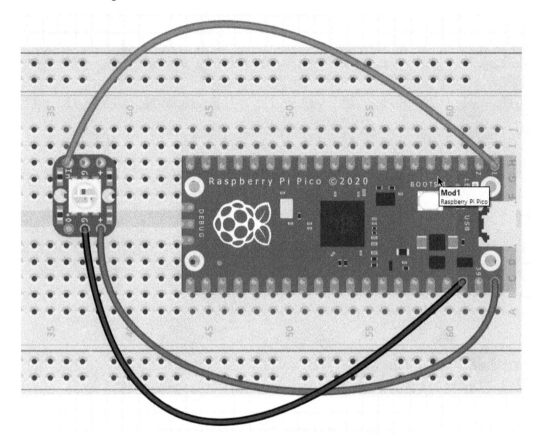

Figure 4.10 – NeoPixel interface to the Pico

Now, let's test if everything is in working order. The first step is to import the requisite modules, as follows:

```
import time
import board
import neopixel
```

Now, let's initialize an object belonging to the Neopixel class by running the following code:

```
pixel_pin = board.GP0
num_pixels = 1
```

```
ORDER = neopixel.RGBW
pixels = neopixel.NeoPixel(
    pixel_pin, num_pixels, brightness=0.2, auto_write=False,
pixel_order=ORDER
)
```

We can create a blinking effect with our NeoPixel, as follows:

```
while True:
    pixels.fill((0, 0, 255, 0))
    pixels.show()
    time.sleep(0.250)
    pixels.fill((0, 0, 0, 0))
    pixels.show()
    time.sleep(0.250)
```

Putting it all together, we have the following:

```
import time
import board
import neopixel

pixel_pin = board.GP0
num_pixels = 1

ORDER = neopixel.RGBW
pixels = neopixel.NeoPixel(
    pixel_pin, num_pixels, brightness=0.2, auto_write=False,
pixel_order=ORDER
)

while True:
    pixels.fill((0, 0, 255, 0))
    pixels.show()
    time.sleep(0.250)
    pixels.fill((0, 0, 0, 0))
    pixels.show()
    time.sleep(0.250)
```

At this point, your project setup should look something like this:

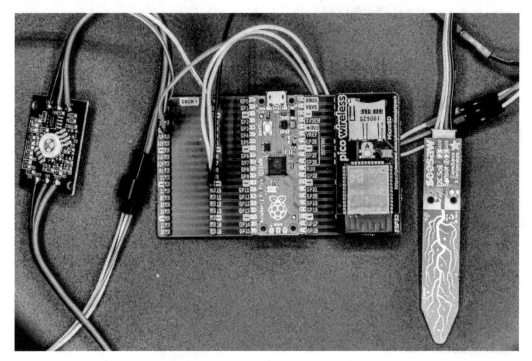

Figure 4.11 – Soil sensor project setup

In the next section, we will discuss publishing the soil moisture level data to **ThingSpeak** and bring our project together.

Publishing data to ThingSpeak

In this section, we will discuss publishing our data to ThingSpeak. If you are not familiar with ThingSpeak, this is an open source software that enables the uploading of data from devices with internet connectivity, such as in our current project. This enables data to be collected from sensors and then analyzed. We chose ThingSpeak as an example of the various toolsets available for developers. Create a free account on `https://thingspeak.com` and create a new channel, as illustrated in the following screenshot:

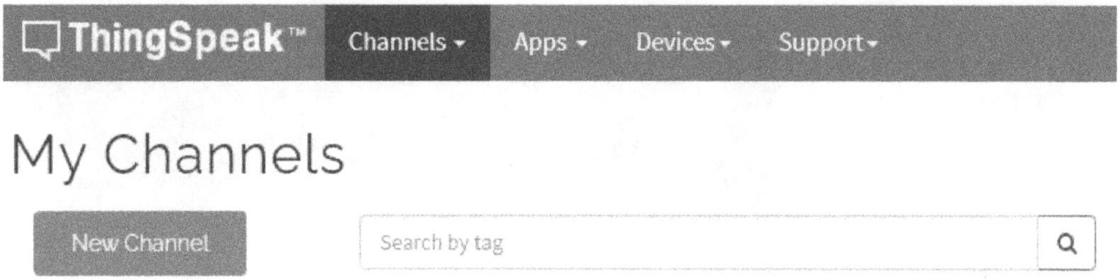

Figure 4.12 – Creating a new channel

Give a name to your channel, fill in **Field 1** (as shown in the following screenshot), and save the channel:

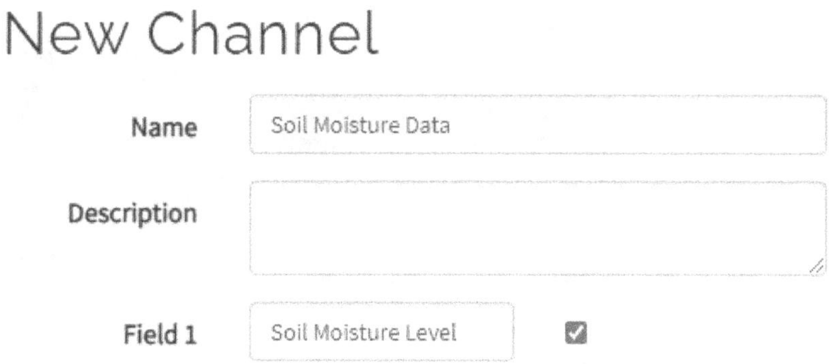

Figure 4.13 – Naming the channel

The channel has now been created. Make a note of the **Write API Key** value from the **API Keys** tab, as illustrated in the following screenshot:

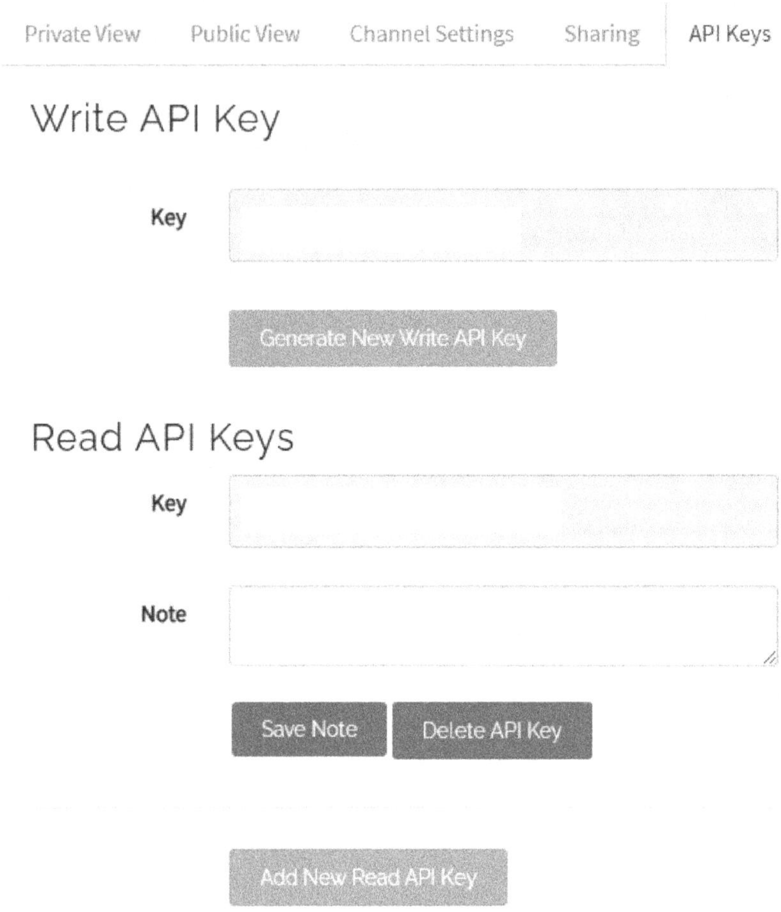

| Private View | Public View | Channel Settings | Sharing | API Keys |

Write API Key

Key

Generate New Write API Key

Read API Keys

Key

Note

Save Note   Delete API Key

Add New Read API Key

Figure 4.14 – Recording the Write API Key value

Now, let's modify `secrets.py` to include the write **application programming interface (API)** key from ThingSpeak, as follows:

```
secrets = {
    'ssid' : 'my_network',
    'password' : 'password',
    'timezone' : "America/New_York",
    'thingspeak_token' : 'ABCD1234',
}
```

The final code sample is available for download along with this chapter as `code_thingspeak.py` (see `https://bit.ly/3avHvy0`). Let's discuss the most important elements of the code, as follows:

- We'll get started by importing the modules required for the project, as illustrated in the following code snippet:

```
import time
import board
import busio
import neopixel
from adafruit_seesaw.seesaw import Seesaw
from digitalio import DigitalInOut
import adafruit_requests as requests
import adafruit_esp32spi.adafruit_esp32spi_socket as
socket
from adafruit_esp32spi import adafruit_esp32spi
```

- We import the credentials needed for the project from `secrets.py`, as follows:

```
try:
    from secrets import secrets
except ImportError:
    print("WiFi secrets are kept in secrets.py, please
add them there!")
    raise
```

- We declare the **Uniform Resource Locator** (**URL**) that is used to publish data to ThingSpeak. The URL is available from the **API Keys** tab of your ThingSpeak account. The code is illustrated here:

```
URL = "https://api.thingspeak.com/update?api_
key={token}&field1={value}"
```

- Now, we initialize the soil sensor that is interfaced via the I2C interface, as follows:

```
i2c = busio.I2C(board.GP9, board.GP8)
ss = Seesaw(i2c, addr=0x36)
```

- Then, we initialize the NeoPixel that is interfaced to GP0, as follows:

```
pixel_pin = board.GP0
num_pixels = 1
ORDER = neopixel.RGBW
pixels = neopixel.NeoPixel(
    pixel_pin, num_pixels, brightness=0.2, auto_
write=False, pixel_order=ORDER
)
```

- Then, we initialize the ESP32 wireless pack. Earlier in this chapter, we discussed the pins used to interface the wireless pack. Here's the code you'll need:

```
esp32_cs = DigitalInOut(board.GP7)
esp32_ready = DigitalInOut(board.GP10)
esp32_reset = DigitalInOut(board.GP11)


spi = busio.SPI(board.GP18, board.GP19, board.GP16)
esp = adafruit_esp32spi.ESP_SPIcontrol(spi, esp32_cs,
esp32_ready, esp32_reset)
```

- Now, we connect it to the wireless network, as follows:

```
requests.set_socket(socket, esp)
if esp.status == adafruit_esp32spi.WL_IDLE_STATUS:
    print("ESP32 found and in idle mode")
print("Firmware vers.", esp.firmware_version)
print("MAC addr:", [hex(i) for i in esp.MAC_address])


print("Connecting to AP...")
while not esp.is_connected:
    try:
        esp.connect_AP(secrets["ssid"],
secrets["password"])
    except RuntimeError as e:
        print("could not connect to AP, retrying: ", e)
        continue
```

```
print("Connected to", str(esp.ssid, "utf-8"), "\tRSSI:",
esp.rssi)
```

```
print("My IP address is", esp.pretty_ip(esp.ip_address))
```

- Now, we enter the main loop of the program. We read the moisture levels every 2 seconds. The code is illustrated in the following snippet:

```
while True:
    # read moisture level
    touch = ss.moisture_read()
    # read temperature from the temperature sensor
    temp = ss.get_temp()
    print("temp: " + str(temp) + "  moisture: " +
str(touch))
```

If the soil moisture level is below a certain threshold, we create a lighting effect on the NeoPixel. This is to remind us that it is time to water the plant. In the following code snippet, the threshold has been set to 500, but this could vary depending upon the type of soil used for your plant. You need to adjust this threshold based upon what you measure by inserting the soil sensor into your potted plant. You also need to measure the wet soil level to set the requisite thresholds:

```
if touch < 500:
        pixels.fill((0, 0, 255, 0))
        pixels.show()
        time.sleep(0.250)
        pixels.fill((0, 0, 0, 0))
        pixels.show()
        time.sleep(0.250)
```

- We also publish the data every time the soil moisture levels fall below the threshold, by running the following code:

```
try:
    r = requests.get(URL.format(token=secrets["thingspeak_
token"], value=str(touch)))
except:
    print("Failed to publish value")
else:
    print("-" * 40)
    print(r.json())
```

```
print("-" * 40)
r.close()
```

We publish the data at a 2-second interval because this will help determine the duration for which the soil moisture levels were below the desired level. Once we have saved the code to the Pico and verified that the data is successfully published to ThingSpeak, it's time to put everything together.

Putting it all together

Once you have adjusted the thresholds, insert the soil sensor next to your potted plant and power up your Pico. You should have something that looks like this:

Figure 4.15 – Fun with gardening

You can also monitor your published data on ThingSpeak, as illustrated in the following screenshot:

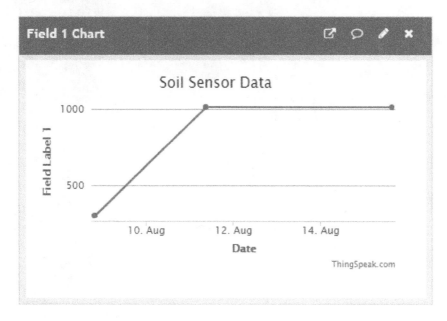

Figure 4.16 – Soil sensor data

ThingSpeak is an analytics platform that lets you aggregate data from your sensors. You can expand this project by creating a dashboard for all the sensors in your garden and remotely control valves from your dashboard. Platforms such as ThingSpeak enable scaling from one sensor to an array of sensors and actuators.

Summary

In this chapter, we discussed interfacing a soil sensor to the Pico and monitoring the moisture levels of potted plants. We also discussed publishing the moisture level to ThingSpeak and triggering an LED when the moisture falls below a certain threshold. This project demonstrates the use of the Pico in gardening activities. You could take this project further by turning on a sprinkler valve when the moisture levels fall below a certain threshold. You could also change the NeoPixel behavior based upon the measured moisture levels—for example, you could make use of a NeoPixel LED strip and make it glow amber when your plant needs watering.

In the next chapter, we will discuss building a weather station using the Pico!

Section 2: Learning by Making

This section is all about gaining experience by building three different projects of varying complexity using the Raspberry Pi Pico. We will start with a weather station and end the section with a visual aid for tracking air quality.

This section contains the following chapters:

5
Building a Weather Station

In this chapter, you will build a weather station to conduct your own citizen science experiments using Raspberry Pi Pico. We will discuss interfacing and testing various sensors with the Pico and installing your Pico outdoors.

By the end of this chapter, you will have built something similar to what is shown in the following figure:

Figure 5.1 – Raspberry Pi Pico interfaced to a weather station

In this project, we will discuss interfacing sensors with the Pico to collect environmental data from the immediate surroundings. We will test all the sensors by discussing the code needed to interface them and put our weather station together.

The topics covered in this chapter will include the following:

- Conducting citizen science experiments
- Installing the requisite libraries
- Testing the sensors
- Assembling and testing the weather station

> **Weatherproofing Your Project**
>
> In this chapter, we want to demonstrate conducting citizen science experiments using the RP2040 microcontroller. We will not discuss power sources or weatherproofing your weather station, as we want to focus on interfacing the sensors and collecting data. Weatherproofing and power source identification require some long-term testing. This project is a work in progress, and we will post our findings to this book's repository.

Technical requirements

The following hardware is recommended for this chapter. We will leave it up to you to select the sensors for your weather station:

- Raspberry Pi Pico (https://bit.ly/3AJtoAf) – $4
- Pico Omnibus – Dual Expander (https://bit.ly/3sr2GJR) – $10.50
- Encased I2C Temperature/Humidity Sensor (https://bit.ly/3FNFWdr) – $29.95
- 2 x RJ11 breakout board (https://bit.ly/3AS6OVQ) – $6.99
- VEML6075 UV Sensor (https://bit.ly/2YYKPQ4) – $7.50
- BME280 Pressure Sensor (https://bit.ly/31Pb3xh) – $14.95
- Adafruit STEMMA Soil Sensor (https://bit.ly/3iQtq3r) – $7.50
- Weather Meter Kit (optional) (https://bit.ly/3johPZC) – $79.95

The code samples discussed in this chapter are available here:

https://github.com/PacktPublishing/Raspberry-Pi-Pico-DIY-Workshop/tree/main/chapter_05

Code in Action videos for this chapter can be viewed at https://bit.ly/3P1uTSz.

Conducting citizen science experiments

Citizen science is a broad approach that involves numerous individuals and groups who volunteer to collect and/or analyze data on various projects that are of wide interest. Participants in these projects may be of any gender, age group, and educational background, and they may span across continents. Some common areas where citizen science can be effective include weather and climate monitoring, wildlife tracking, and related areas. Applications include general science, sustainability, climate change, population health, and smart cities. It is expected that the nature, scope, and number of citizen science endeavors will grow immensely in years to come. The types of data gathered by citizen science may include discrete or continuous parameters such as temperature, **Volatile Organic Compounds** (**VOCs**), and other atmospheric content, images, and audio files.

At scale, citizen science projects can be effective at helping solve large-scale data-intensive problems in an inexpensive manner. When citizen science projects require hardware, scaling can become prohibitively expensive if care is not taken to minimize unit costs. If care is taken to choose inexpensive components such as the Raspberry Pi Pico, the projects can grow in scope and reach a wide gamut of participants. The weather station demonstrated in this chapter can serve as an example of how to create low-cost data-driven citizen science solutions.

Installing the requisite libraries

We are going to install all the libraries required for this chapter. The first step is to download the CircuitPython Library Bundle. You can download the Library Bundle as a ZIP file from here: `https://circuitpython.org/libraries`.

Extract the contents of the ZIP file to get started with the library installation.

> **CircuitPython Installation**
>
> We are assuming that you have installed CircuitPython on your Pico. If you are not familiar with the installation process, we recommend following the installation process from *Chapter 1, Getting Started with Raspberry Pi Pico*.

The AM2315 sensor

We need two libraries for the AM2315 temperature (shown in the following figure) and humidity sensor, namely `adafruit_bus_device` and `adafruit_ahtx0`. In the extracted bundle, there is a folder called `adafruit_bus_device` along with the `adafruit_ahtx0.mpy` file. Copy over the folder to the `lib` folder of the Pico.

> **Library Selection**
>
> According to the guide provided by Adafruit, you can read temperature and humidity from the AM2315 sensor using the `adafruit_am2320` library (`https://bit.ly/3psAUgE`). However, we had problems with initialization while using the library. Since the I2C address of the sensor was identical to that of an AHT20, we decided to try the `adafruit_ahtx0` library, and we were able to read the sensor data.

BME280 sensor

We need two libraries for the BME280 sensor, namely `adafruit_bus_device` and `adafruit_bme280`. Since we already copied over the former in the previous section, copy the latter of the same name to the `lib` folder of the Pico.

VEML6075 UV light sensor

We need two libraries for the VEML6075 sensor, namely `adafruit_bus_device` and `adafruit_veml6075`. Since we already copied over the former, we need to copy over `adafruit_veml6075.mpy` to the `lib` folder of the Pico.

Testing the sensors

In this section, we will set up and test the individual components used in the project before we put them together.

Testing the BME280 sensor

In this section, we will test the BME280 sensor (as shown in the following screenshot):

Figure 5.2 – The BME280 Pressure Sensor

The BME280 Pressure Sensor can be used to measure temperature, humidity, and atmospheric pressure. It comes with an I2C interface, and its address on the I2C bus is 0x77.

The sensor is interfaced to the Pico, as shown in the following schematic:

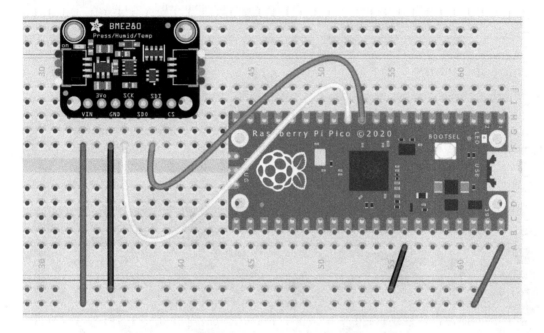

Figure 5.3 – The Fritzing schematic for the BME280 sensor

In the preceding schematic, the sensor is interfaced to the Pico as follows, where the left side of the arrow refers to a pin on the BME280 breakout board, while the right side of the arrow refers to a pin on the Raspberry Pi Pico:

- VIN → VBUS
- SCL → GP9
- SDA → GP8
- GND

> **Pull-Up Resistors for I2C Bus**
>
> Typically, pull-up resistors are needed for the SCL and SDA pins of the I2C bus. The STEMMA soil sensor comes with pull-up resistors on the I2C bus. While using off-the-shelf sensors that come with an I2C interface, read the schematic and understand the I2C bus configuration of the sensor. If you are not familiar with the need for pull-up resistors for the I2C bus, we recommend checking out *Chapter 2, Serial Interfaces and Applications*.

Qwiic connectors

The BME280 sensor comes with a pair of Qwiic/STEMMA connectors. If you have a Qwiic jumper cable (Link: `https://www.sparkfun.com/products/17261`), you can connect it, as shown in the following figure. If you are not familiar with the Qwiic/STEMMA ecosystem, we recommend reading *Chapter 2, Serial Interfaces and Applications*. Qwiic connectors make prototyping easier by daisy-chaining the breakout boards:

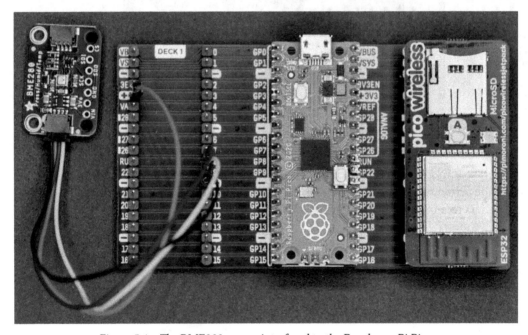

Figure 5.4 – The BME280 sensor interfaced to the Raspberry Pi Pico

Let's review the code needed to interface the sensor. The code sample discussed is available for download with this chapter as `code_bme280.py`:

1. The first step to reading data from the sensor is to import the requisite modules. The `adafruit_bme280` library comes with two submodules, namely `basic` and `advanced`. We are going to be using the `basic` module in this example. For the sake of readability, we are going to import `adafruit_bme280.basic` as `adafruit_bme280`:

```
import board
import busio
import time
from adafruit_bme280 import basic as adafruit_bme280
```

2. The next step is to initialize the I2C bus and pass the bus object as an argument to the `BME280` class:

```
i2c = busio.I2C(board.GP9, board.GP8)
bme280 = adafruit_bme280.Adafruit_BME280_I2C(i2c)
```

3. Use your local weather report to set your local pressure at sea level:

```
bme280.sea_level_pressure = 1007.1
```

4. Now, we can read the temperature, pressure, and humidity. In the following code snippet, you will note that the text is split across two lines. There are actually no line breaks, and the text formatting splits it across two lines:

```
while True:
    print("\nTemperature: %0.1f C" % bme280.temperature)
    print("Humidity: %0.1f %%" % bme280.relative_
humidity)
    print("Pressure: %0.1f hPa" % bme280.pressure)
    print("Altitude = %0.2f meters" % bme280.altitude)
    time.sleep(2)
```

5. Upon executing the code, you should see output from the sensor, as follows:

```
CircuitPython REPL

Pressure: 986.2 hPa
Altitude = 176.36 meters

Temperature: 23.8 C
Humidity: 57.9 %
Pressure: 986.2 hPa
Altitude = 176.21 meters

Temperature: 23.8 C
Humidity: 57.9 %
Pressure: 986.2 hPa
Altitude = 176.51 meters
```

Figure 5.5 – BME280 Sensor output seen in Mu IDE

Now that we have tested the BME280 sensor, we will discuss testing the AM2315 sensor in the next section.

Testing the AM2315 temperature/humidity sensor

In this section, we will test the AM2315 temperature/humidity sensor (as shown in the following figure):

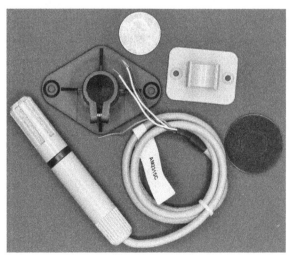

Figure 5.6 – The AM2315 sensor

Since the AM2315 sensor comes with an I2C interface, the connections are identical to that of the BME280 sensor (refer to the schematic shown in *Figure 5.5*).

Let's take a quick look at the code needed to interface the sensor. The code sample discussed in this section is available for download along with this chapter as `code_ahtx0.py`:

1. As always, let's get started with testing the sensor by importing the requisite modules:

```
import time
import board
import busio
import adafruit_ahtx0
```

2. We initialize the I2C bus and pass the object to the AHTx0 class. As discussed earlier, we are using the `adafruit_ahtx0` library because the temperature sensor chip used is compatible with the `ahtx0` family:

```
i2c = busio.I2C(board.GP9, board.GP8)
temp_sensor = adafruit_ahtx0.AHTx0(i2c)
```

3. Now, it is time to read the temperature and humidity values. In the following code snippet, you will note that the text is split across two lines. There are no line breaks, and the text formatting splits it across two lines:

```
while True:
    print("\nTemperature: %0.1f C" % sensor.temperature)
    print("Humidity: %0.1f %%" % sensor.relative_
humidity)
    time.sleep(2)
```

4. We should be able to see the temperature and humidity values on the serial port output when we save the preceding code sample as code.py:

```
CircuitPython REPL
Temperature: 22.1 C
Humidity: 55.2 %

Temperature: 22.1 C
Humidity: 55.2 %

Temperature: 22.1 C
Humidity: 55.2 %

Temperature: 22.1 C
Humidity: 55.2 %
```

Figure 5.7 – The temperature and humidity values

In the next section, we will discuss testing the VEML6075 UV sensor.

Testing the VEML6075 sensor

In this section, we will calculate the UV index using the VEML6075 sensor (as shown in the following figure). The libraries we installed earlier in this chapter provide a UV index output using the raw sensor values:

Figure 5.8 – The VEML6075 sensor

The code sample discussed in this section is available for download along with this chapter as code_veml6075.py. The interfacing of the VEML6075 sensor is identical to that of the BME280 sensor.

Let's take a look at the code needed to interface the sensor:

1. As always, let's get started with testing the sensor by importing the requisite modules:

```
import time
import board
import busio
import adafruit_veml6075
```

2. We initialize the I2C bus and pass the object to the VEML6075 class:

```
i2c = busio.I2C(board.GP9, board.GP8)
veml = adafruit_veml6075.VEML6075(i2c, integration_
time=100)
```

3. Now, it is time to read the UV index values:

```
while True:
    print(veml.uv_index)
    time.sleep(1)
```

4. You can test the sensor by using a UV flashlight or exposing it to sunlight. In the following screenshot, you can see the spike in UV index values when the sensor is exposed to a UV flashlight:

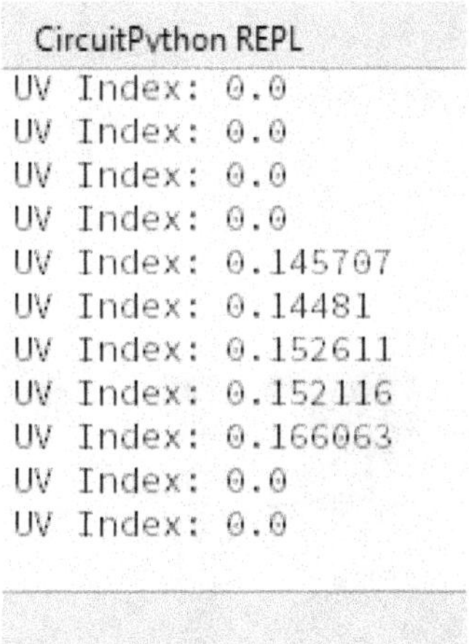

```
CircuitPython REPL
UV Index: 0.0
UV Index: 0.0
UV Index: 0.0
UV Index: 0.0
UV Index: 0.145707
UV Index: 0.14481
UV Index: 0.152611
UV Index: 0.152116
UV Index: 0.166063
UV Index: 0.0
UV Index: 0.0
```

Figure 5.9 – The spike in UV index values using a UV flashlight

Caution
UV flashlights are not toys; do not look at the light directly.

In the next section, we will test all the sensors belonging to the weather meter.

Testing the weather meter sensors

In this section, we will discuss testing the sensors belonging to the weather meter kit, namely the rainfall sensor, anemometer, and wind vane.

> **Weather Meter Selection**
>
> The weather meter that we chose is quite expensive for a DIY project. It is possible to create your own anemometer, wind vane, and so on via 3D printing. Alternatively, you can build a desktop weather station that makes use of public weather sources to retrieve weather alerts and so on.

Testing the rainfall sensor

As the name indicates, the rainfall sensor (shown in the following figure) is used to detect rainfall. The rainfall sensor is a tipping bucket with a tilt switch. According to the data sheet, the switch closes for every *0.011 inches* of rainfall. The bucket tips over and opens the switch. It basically acts as a momentary press switch. This enables the calculation of total rainfall over a given window:

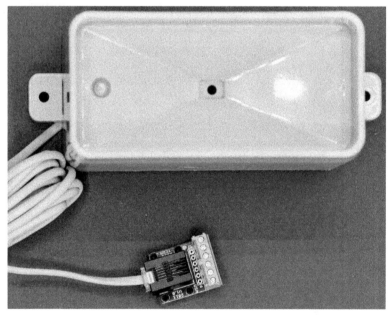

Figure 5.10 – The rainfall sensor connected to the RJ11 breakout board

The rainfall sensor is interfaced to the Pico, as shown in the following figure:

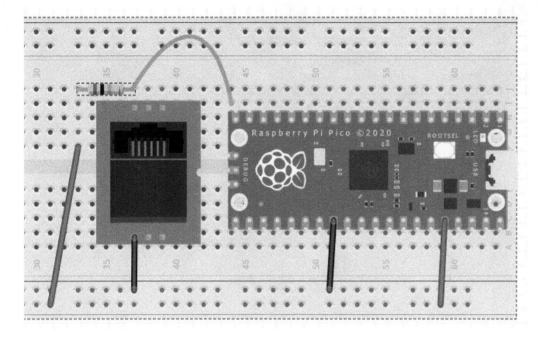

Figure 5.11 – The rainfall sensor schematic

According to the data sheet, the tilt switch of the rainfall sensor is connected to the third and fourth pins of the RJ11 connector. We tie one end of the switch to the ground, while the other end is connected to GPIO pin 15 and pulled up using a 10K resistor. Now, we will discuss the code needed to detect rainfall.

We will be making use of the countio module to count the pin state transitions. The countio module counts any falling edge. Since we have pulled up GPIO pin 15 and one end of the switch to 3.3V, the pin is pulled down every time the switch closes. The code sample discussed in this chapter is available for download along with this chapter as code_rainfall.py:

1. We start by importing all the requisite modules:

```
import time
import board
import countio
```

2. We initialize Counter on pin GP15 (GPIO pin 15) and start counting the pulses:

```
with countio.Counter(board.GP15) as pin_counter:
    while True:
        if pin_counter.count:
```

```
print("Rainfall Detected")
pin_counter.reset()
time.sleep(2)
```

3. Save the preceding sample as code.py and introduce a couple of droplets of water. You should see the following output:

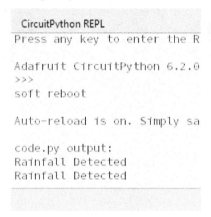

Figure 5.12 – The rainfall sensor

In the next section, we will test the anemometer and the wind vane.

Testing the anemometer and the wind vane

In this section, we will test the anemometer and the wind vane. Both the sensors are connected to the same RJ11 connector (as shown in the following figure):

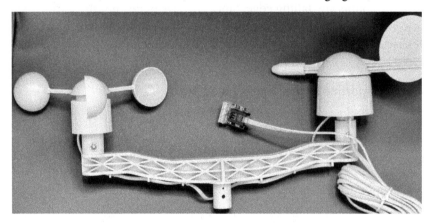

Figure 5.13 – The anemometer and wind vane connected to the RJ11 breakout board

Let's look at how to read the wind speed and direction.

Anemometer

An anemometer is an instrument (shown on the left side of the previous figure) that is used to measure wind speed. According to the data sheet, the anemometer comes with a reed switch. When the wind moves the rotating element, the magnet closes the switch. Every switch closure corresponds to `1.492` mph of wind speed. The schematic needed to read the wind speed is identical to that of *Figure 5.13*, except that the anemometer pin is connected to `GPIO pin 14`.

Let's take a quick look at the code needed to measure wind speed. The code sample discussed in this section is available as `code_anemometer.py`:

1. The first step is importing all the requisite modules. As with the previous example, we are going to be using the `countio` module:

    ```
    import time
    import board
    import countio
    ```

2. We initialize a counter on `pin 14` and count the pulses. Since a pulse amounts to 1.492 mph of wind speed, we count the number of pulses in a second using `time.monotonic()` because it returns the time in seconds. We can make use of it to calculate wind speed. In the following code snippet, you will note that the text is split across two lines. There are no line breaks, and the text formatting splits it across two lines:

    ```
    with countio.Counter(board.GP14) as pin_counter:
        last_tick = time.monotonic()
        while True:
            if pin_counter.count and ((time.monotonic() -
    last_tick) > 1.0):
                print("Wind Speed = {0} mph".format(pin_
    counter.count * 1.492))
                pin_counter.reset()
                last_tick = time.monotonic()
    ```

3. Save the code sample as code.py and try spinning the anemometer; you should start seeing wind speed output, as shown in the following figure:

```
CircuitPython REPL
Auto-reload is on. Simply save

code.py output:
Wind Speed = 0.535697 mph
Wind Speed = 4.47163 mph
Wind Speed = 0.504894 mph
Wind Speed = 4.47163 mph
Wind Speed = 0.885686 mph
```

Figure 5.14 – The wind speed calculation

We can verify the wind speed by installing the weather meter kit outside and comparing it with data from public data sources. In the next section, we are going to test the wind vane.

Testing the wind vane sensor

In this section, we are going to test the wind vane. A wind vane is an instrument used to determine wind direction. The wind vane also uses reed switches in combination with resistors to determine the direction. We can make use of a potential divider to read the wind direction.

The wind vane is interfaced to the Pico, as shown in the following schematic. We will connect the potential divider to GPIO pin 26:

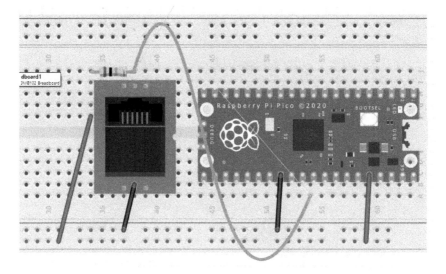

Figure 5.15 – Interfacing the wind vane to the Pico

The code sample discussed in this section is `code_windvane.py`. Let's take a look at it:

1. The first step is to download the requisite modules. Since the wind vane is an analog output, we will import the `analogio` module:

```
import time
import board
import analogio
```

2. We initialize `GP26` as an **Analog to Digital Converter (ADC)** input:

```
vane = analogio.AnalogIn(board.GP26)
```

3. Then, we read the ADC value, convert into voltage, and calculate the direction:

```
while True:
    raw = vane.value
    volts = get_voltage(raw)
    print("raw = {:5d} volts = {:5.2f} angle={}".
format(raw, volts, wind_dir(volts * 1000)))
    time.sleep(0.5)
```

4. The direction is calculated based upon the resistance values provided in the data sheet. We make use of a 10K resistor to create a potential divider:

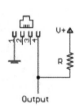

Example wind vane interface circuit. Voltage readings for a 5 volt supply and a resistor value of 10k ohms are given in the table.

Direction (Degrees)	Resistance (Ohms)	Voltage (V=5v, R=10k)
0	33k	3.84v
22.5	6.57k	1.98v
45	8.2k	2.25v
67.5	891	0.41v
90	1k	0.45v
112.5	688	0.32v
135	2.2k	0.90v
157.5	1.41k	0.62v
180	3.9k	1.40v
202.5	3.14k	1.19v
225	16k	3.08v
247.5	14.12k	2.93v
270	120k	4.62v
292.5	42.12k	4.04v
315	64.9k	4.78v
337.5	21.88k	3.43v

Figure 5.16 – The direction versus resistance values from the wind vane data sheet

5. The RP2040's ADC resolution is 12 bits, but the CircuitPython Library scales it to 16 bits. The ADC outputs a value between 0 and 65535, where 0 represents 0V and 65535 represents 3.3V. Hence, we use a helper function called get_voltage to calculate the voltage from the raw ADC values:

```
def get_voltage(raw):
    return (raw * 3.3) / 65536
```

6. Then, we use the calculated voltage to determine the wind direction (using information from the data sheet):

```
def wind_dir(vin):
    if vin < 150: windDir="202.5"
    elif vin < 300: windDir = "180"
    elif vin < 400: windDir = "247.5"
    elif vin < 600: windDir = "225"
    elif vin < 900: windDir = "292.5"
    elif vin < 1100: windDir = "270"
    elif vin < 1500: windDir = "112.5"
    elif vin < 1700: windDir = "135"
    elif vin < 2250: windDir = "337.5"
    elif vin < 2350: windDir = "315"
    elif vin < 2700: windDir = "67.5"
    elif vin < 3000: windDir = "90"
    elif vin < 3200: windDir = "22.5"
    elif vin < 3400: windDir = "45"
    elif vin < 4000: windDir = "0"
    else: windDir = "0"

    return windDir
```

7. When we save the code sample as code.py, we should see the following output when we try to move the wind vane around:

```
CircuitPython REPL
raw = 29232 volts =  1.47 angle=112.5
raw = 29136 volts =  1.47 angle=112.5
raw = 28800 volts =  1.45 angle=112.5
raw = 28544 volts =  1.44 angle=112.5
raw = 48448 volts =  2.44 angle=67.5
raw = 48800 volts =  2.46 angle=67.5
```

Figure 5.17 – Wind direction values

Now that we have tested the weather meter sensors, it is time to put everything together.

Testing the wireless pack

The wireless pack can be used to upload data to the cloud or retrieve weather information. We recommend checking out *Chapter 2, Serial Interfaces and Applications*, to set up and test the wireless pack.

Assembling and testing the weather station

We recommend following an excellent guide from SparkFun to assemble the kit. You can refer to the guide here: `https://bit.ly/3yTgZZu`. We assembled our kit and tried setting it up in our backyard, as shown in the following figure. We made use of a Pico and a weather carrier board for our installation. This is just for the sake of convenience to avoid some messy wires outdoors:

Figure 5.18 – A weather station installation in the backyard

We tested our weather station without the following considerations:

- A power source for the weather station

- Assembly of the carrier board onto the weather station

- Weather-proofing the carrier board inside an enclosure

- Sturdy installation of the weather station

It is going to take some extensive testing to identify and fix all the problems. The first step is to fix any problems associated with sensor interfacing. Then, we need to ensure that any enclosure we choose for the carrier board can withstand winter weather conditions, including high winds and snowfall. The power source also needs to be able to endure such conditions. There is an interesting writeup from SparkFun about lessons learned from their trial deployment: `https://learn.sparkfun.com/tutorials/weather-station-wirelessly-connected-to-wunderground#lessons-learned`.

In the next section, we will discuss the next steps we intend to take with this project.

Next steps

At the time of writing, we did not come across any add-on hardware board for the Pico that specifically accommodates weather meter sensors. We plan to spin our own **Printed Circuit Board** (**PCB**) to avoid any messy wires. We will share our progress on this book's GitHub repository.

We would like to point out that this project is a work in progress, and we will publish any modifications to the project, including any shortcomings that we might have missed during the build.

Summary

In this project, we discussed conducting our own citizen science experiment by building our own weather station using the Raspberry Pi Pico. We tested the sensors and put the weather station together. We also discussed troubleshooting problems associated with deploying the station outdoors, and we also discussed weather-proofing our installation.

Since citizen science experiments benefit from sharing data on publicly available data streams, we plan to make our data available once we have added connectivity to our weather station. We also plan to publish the status of our weather station to a public data stream.

In the next chapter, we are going to have fun with displays!

6

Designing a Giant Seven-Segment Display

In this chapter, we will have fun with some retro-style displays. We will build a giant display driven by the Pico. It is a fun experience to build visual aids involving LEDs and seven-segment displays. By the end of this chapter, you should be able to build something like the display shown in the following figure, where we have a seven-segment display inside a shadow box:

Figure 6.1 – Giant seven-segment display

In this chapter, we will discuss building such visual aids. We will also discuss options to build displays of different sizes and applications. The topics discussed in this chapter include the following:

- Inspiration for the project
- Installing the required libraries
- Selecting a seven-segment display
- Wiring up the giant seven-segment display
- Writing the drivers for the giant seven-segment display
- Using the display
- Putting it all together

Technical requirements

The following hardware are recommended for this chapter:

- Raspberry Pi Pico (link: `https://bit.ly/3AJtoAf`) – USD 4
- Pico Omnibus – Dual Expander (link: `https://bit.ly/3sr2GJR`) – USD 10.50
- Pico Wireless Pack (link: `https://bit.ly/3yPuoT9`) – USD 16.75
- 6.5" seven-segment display (link: `https://bit.ly/3z6ii76`) – USD 18.95
- Seven-segment digit driver (link: `https://bit.ly/3hpLC2F`) – USD 8.50
- Jumper wires (link: `https://amzn.to/3nmRI7L`) – USD 5.99
- 12 V power supply (link: `https://amzn.to/3q9wcmB`) – USD 8.89
- Shadow box from your local hobby store (something like the box shown in the previous figure)

The code samples for this chapter are available from `https://github.com/PacktPublishing/Raspberry-Pi-Pico-DIY-Workshop/tree/main/chapter_06`.

Code in Action videos for this chapter can be viewed at `https://bit.ly/3MN01BO`.

Seven-Segment Digits

For this project, we have used five seven-segment displays to display any number up to five digits. The code discussed in the chapter can be easily adapted to the number of digits needed for your project.

In the next section, we will discuss the inspiration for the project in this chapter.

Inspiration for the project

We have a general interest in building visual aids for improving quality of life. These visual aids can be used to transmit calculated data and serve to inform users on progress toward goals, suggest how much progress is remaining, or for other purposes. We have used this in healthcare settings previously. For instance, people with chronic health conditions, or a propensity to develop chronic conditions such as diabetes or obesity, can greatly benefit from such displays. In the past, we have developed solutions that recorded daily step counts accomplished, as well as the opposite, the remainder of the steps required to meet daily goals, which, when displayed through these visual aids, and when displayed prominently, can be unmissable.

Using the Pico, such visual aids can be developed for similar applications as well as for uses far and wide. For instance, visual aids can be used to track scores between teams competing as motivation to accomplish health goals. Similarly, alphanumeric visual aids can be built for applications outside the healthcare area as well.

For our previous book, *Python Programming with Raspberry Pi*, we built a Raspberry Pi Zero-based personal health dashboard that tracked physical activity in the form of a progress bar. The dashboard consisted of eight RGB LEDs and an LED was lit green for every 1,250 steps tracked using a fitness tracker. The following figure shows the Raspberry Pi Zero-based *progress bar-like* dashboard:

Figure 6.2 – Raspberry Pi Zero-based personal health dashboard

Since the dashboard only keeps track of the progress of physical activity and does not provide any information on the actual number of steps left for the day, we decided to build a bigger dashboard. This would enable planning the appropriate physical activity based upon the time available. Hence, we built the giant dashboard. We decided to share our work at Maker Faires, as shown in the following figure, where we displayed our work at the Cleveland Maker Faire:

Figure 6.3 – Giant display exhibit at the Cleveland Maker Faire

The preceding figure shows both the dashboards right next to each other. You will have noticed that the giant display shows the number *7161*. It indicates that the person wearing the fitness tracker has *7161* steps left to complete their daily step goal. Instead of counting up to the daily step goal, the dashboard counts down from the goal. This enables planning your physical activity according to the steps left for the day.

We decided to re-design this display using the Pico and wanted to share our design in this chapter.

Potential use cases

Some potential use cases for the project include the following:

- Tracking physical activity

- Keeping track of a head count in a space

- Keeping track of followers on social media platforms

- Keeping track of scores for contests

In this chapter, we will show how to build a display for any of these purposes.

Installing the required libraries

In this section, we will install the libraries required for the project. These include the drivers for the wireless pack and the seven-segment display.

> **CircuitPython Installation**
>
> We are assuming that you have installed CircuitPython on your Pico. If you are not familiar with the installation process, we recommend following the installation process from *Chapter 1, Getting Started with the Raspberry Pi Pico*.

The libraries are all part of the Adafruit CircuitPython bundle. The latest bundle could be downloaded as a ZIP file from `https://circuitpython.org/libraries`. We used the bundle version meant for CircuitPython 6.x.x.

After downloading the ZIP file, extract its contents so that you can copy the libraries we need for the project.

Wireless pack

We need the `adafruit_esp32spi` library for the wireless pack. Copy over the folder (with the same name) to the `lib` folder of your Pico. We will also need the `adafruit_reqests.mpy` binary from the bundle.

> **Setting Up the Wireless Pack**
>
> In *Chapter 2, Serial Interfaces and Applications*, we discussed setting up the wireless pack in detail. We recommend reading through it and setting up your wireless pack.

Selecting a seven-segment display

We chose a 6.5" seven-segment display for this chapter. We chose it because of its giant size but each digit needs a separate driver. The following figure shows a seven-segment digit with a driver soldered on its back:

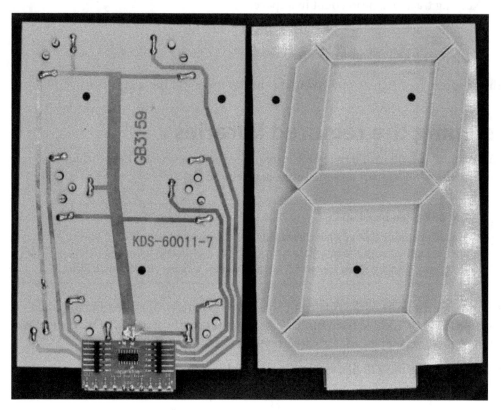

Figure 6.4 – 6.5" seven-segment digit with a driver on the back

The total cost of a single digit comes to around **USD 30**. It can get quite expensive and so we wanted to suggest some alternatives that might suit your budget. We found a smaller display on eBay for **USD 4**. It is available from here: `https://ebay.to/3nKtd4F`.

You should be able to use the drivers written for the giant seven-segment display to drive this tiny display (shown in the following figure). The only disadvantage of this display is that it is very small and the numbers cannot be seen from a distance:

Figure 6.5 – Small seven-segment display

If you are looking for a slightly bigger display at an additional cost, we recommend the seven-segment backpack from Adafruit Industries. It comes with an I2C interface, thus making it easy to wire it up to the Pico. Adafruit also provides drivers for this display:

Figure 6.6 – HT16K33 seven-segment backpack from Adafruit Industries

Since we are fans of the giant seven-segment display from SparkFun (shown in *Figure 6.4*), we chose it to build our giant display. In the next section, we will set up the wireless pack to connect to the local wireless network.

Wiring up the giant seven-segment display

In this section, we will wire up the display to the Raspberry Pi Pico. We used the **Pico Omnibus – Dual Expander** from Pimoroni. This enables interfacing the wireless pack and wiring up the seven-segment display. The steps to interface the display are as follows:

1. Assemble the individual seven-segment digits. The driver needs to be soldered onto the back of each digit. Soldering the driver is a very simple step and we followed the instructions available from SparkFun (link: `https://bit.ly/3hLUobk`).

2. Then, we connected the seven-segment driver to the following pins of the Pico, as shown in the following figure:

    - Latch | GP17

    - Clock | GP18

    - Serial | GP19

    - 5V | VBUS

    - 12V | External power supply

    The following figure shows the Fritzing schematic for the connections between the seven-segment driver and the Raspberry Pi Pico:

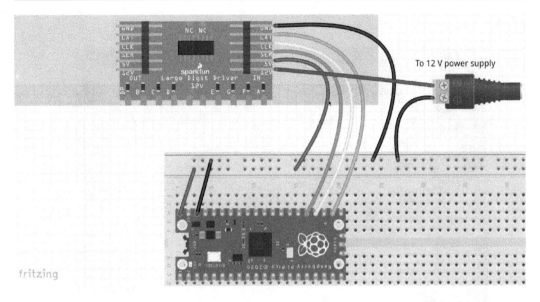

Figure 6.7 – Fritzing schematic to interface the Pico to the seven-segment driver

3. The driver comes with two sets of pins, namely **In** and **Out**. We connected the Pico to the **In** side.

4. We also ensured that the grounds of the external power supply and the Pico were tied together. The wiring setup is shown in the previous figure.

5. The next step is to serially connect as many digits as you want. The **Out** set of pins is connected to the **In** side of the next digit, as shown in the following Fritzing schematic:

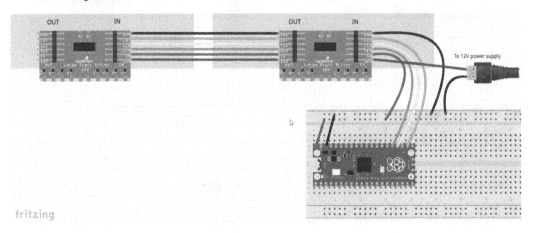

Figure 6.8 – Two seven-segment drivers connected in series

6.   We connected up to five digits for our project:

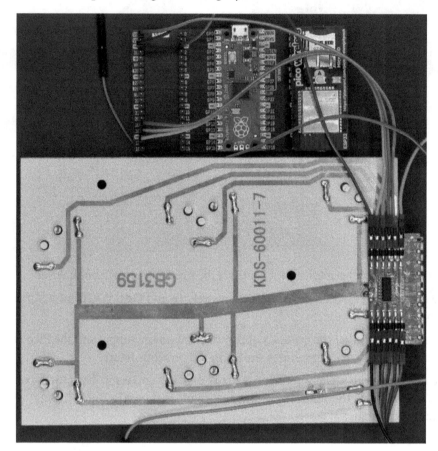

Figure 6.9 – Giant seven-segment display interfaced to a Raspberry Pi Pico

In the next section, we will discuss the drivers for the display.

Writing the drivers for the giant seven-segment display

In the previous section, we wired up the display. Now, we need to write the drivers to test the display. We ported the C++ code provided by SparkFun to CircuitPython:

1.   The first step is to import all the required modules for the driver:

```
import board
import busio
```

```
import time
from digitalio import DigitalInOut, Direction, Pull
```

2. We declare the latch, clock, and data pins and set them up as output pins. We also set them all to *low*:

```
latch = DigitalInOut(board.GP22)
clock = DigitalInOut(board.GP26)
data = DigitalInOut(board.GP27)

latch.direction = Direction.OUTPUT
clock.direction = Direction.OUTPUT
data.direction = Direction.OUTPUT

latch.value = False
clock.value = False
data.value = False
```

3. The post_number function is used to display a number, decimal place, and so on. This is accomplished by turning on the corresponding segments of the digit. For example, in order to display the number 1, we need to turn on the **B** and **C** segments. The following figure shows the segment names of a seven-segment display:

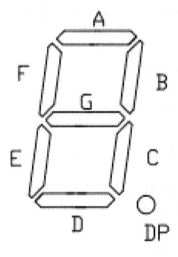

Figure 6.10 – Segment names of a seven-segment digit

4. The driver board consists of an 8-bit shift register. The number to be displayed is clocked out through the *serial* pin:

```python
def post_number(number, decimal=False):
    a = 1<<0
    b = 1<<6
    c = 1<<5
    d = 1<<4
    e = 1<<3
    f = 1<<1
    g = 1<<2
    dp = 1<<7

    if number == 1: segments = b | c
    elif number == 2: segments = a | b | d | e | g
    elif number == 3: segments = a | b | c | d | g
    elif number == 4: segments = f | g | b | c
    elif number == 5: segments = a | f | g | c | d
    elif number == 6: segments = a | f | g | e | c | d
    elif number == 7: segments = a | b | c
    elif number == 8: segments = a | b | c | d | e | f | g
    elif number == 9: segments = a | b | c | d | f | g
    elif number == 0: segments = a | b | c | d | e | f
    elif number == " ": segments = 0
    elif number == "c": segments = g | e | d
    elif number == "-": segments = g
    else: segments = 0

    if decimal: segments |= dp

    for i in range(8):
        clock.value = False
        data.value = segments & (1 << (7 - i))
        clock.value = True
```

5. As mentioned earlier, the digits are serially connected to each other. Since we are using five digits in this project, we can display numbers at each position by shifting the digits. The show_number function is used to extract the digits from the number and display it:

```
def show_number(x):

    for num in range(5):
        remainder = int(x % 10)
        post_number(remainder, False)
        x = int(x / 10)

    latch.value = False
    latch.value = True
```

Since our display consists of five digits, we repeat the for loop five times. This ensures that we add preceding zeroes to the number if necessary. The following figure shows two digits of the display showing the number *47* during testing:

Figure 6.11 – Display interfaced to a Pico

The driver we discussed in this section is available for download along with the rest of this chapter's materials as `seven_segment.py`. Save the file to your Pico as we will be using it in our examples. In the next section, we will show how to put everything together.

Using the display

In this section, we will discuss driving the displays in two ways: with a simple web server running on the Pico and via a serial port.

Let's get started!

Simple web server

In this example, we will host a simple web server on the Pico so that we can update the display using a browser from any device on a local network. This example is based on the `esp32spi_wsgiserver` example made available from Adafruit (link: `https://bit.ly/3itHrmY`). We made some simple modifications to the code sample to adapt it to our example. The modified code sample is available for download from this chapter's repository as `code_server.py` (link: `https://bit.ly/3hqWA7X`).

The changes made to the server example include the following:

- Importing the `seven_segment` driver so that we can update the display when there is a request:

```
import seven_segment
```

- We updated the GPIO pin numbers to drive the ESP32 wireless pack:

```
# If you have an externally connected ESP32:
esp32_cs = DigitalInOut(board.GP7)
esp32_ready = DigitalInOut(board.GP10)
esp32_reset = DigitalInOut(board.GP11)


spi = busio.SPI(board.GP18, board.GP19, board.GP16)
esp = adafruit_esp32spi.ESP_SPIcontrol(
    spi, esp32_cs, esp32_ready, esp32_reset
)
```

- We registered a method called `display` to update the display when there is an incoming GET request:

```
web_app = SimpleWSGIApplication(static_dir=static)
web_app.on("GET", "/display", display)
```

- The `display` method is called whenever the `display` path is called in the request. The `display` method includes the following:

```
def display(environ):  # pylint: disable=unused-argument
    if "QUERY_STRING" in environ.keys():
        query = environ['QUERY_STRING']
        if "value=" in query:
            value = query[6:]
            print(value)
            seven_segment.show_number(int(value))
    return web_app.serve_file("static/index.html")
```

- In the `display` method, we check whether the request includes any query parameters. If there is a parameter named `value`, we parse and convert it into an integer.

- Then, we use the `seven_segment` driver to update the display.

- We also need to return a response. The method returns a static web page that reads `Success!`. The static HTML file is available along with this chapter's code samples.

- It is time to take the server for a spin. Once you have wired up the display, your Pico needs to contain the following files, as shown in *Figure 6.12*:

  - The seven-segment driver for the display (`seven_segment.py`)

  - `secrets.py` for the Wi-Fi credentials

  - HTML files under the `static` folder

The files are as shown in the following screenshot:

Name	Date modified	Type	Size
.fseventsd	12/31/2019 11:00 PM	File folder	
lib	12/31/2019 11:00 PM	File folder	
static	9/9/2021 12:09 AM	File folder	
.metadata_never_index	12/31/2019 11:00 PM	METADATA_NEVE...	0 KB
.Trashes	12/31/2019 11:00 PM	TRASHES File	0 KB
boot_out.txt	12/31/2019 11:00 PM	Text Document	1 KB
code.py	10/3/2021 5:01 PM	Python File	7 KB
secrets.py	9/9/2021 12:08 AM	Python File	1 KB
seven_segment.py	10/3/2021 1:56 PM	Python File	2 KB

Figure 6.12 – Files for the web server example

- Once you have loaded the code sample as code.py and opened a serial terminal, you will notice that the Pico makes use of the credentials provided in secrets.py to connect to your wireless network. It also starts a server and provides the IP address of the Pico, as shown in the following screenshot:

```
code.py output:
ESP32 SPI simple web server test!
MAC addr: ['0x0', '0x65', '0x9a', '0x57', '0xdd', '0xc4']
MAC addr actual: ['0xc4', '0xdd', '0x57', '0x9a', '0x65', '0x0']
open this IP in your browser:  192.168.86.32
```

Figure 6.13 – Pico web server IP address

- We will make use of the Pico's IP address to update the display. As shown in *Figure 6.13*, the Pico's IP address is 192.168.86.32. Let's make a GET request to display the number 789.

- Launch a browser from a device connected to the same network as your Pico. Enter the following in the address bar: http://192.168.86.32/display?value=789.

- It should return a static web page (as shown in *Figure 6.14*) if the GET request was successful:

← → C ⚠ Not secure | 192.168.86.32/display?value=789

Success!

Figure 6.14 – Pico web server response

- Your display should have been updated to 00789 (notice the preceding zeroes), as shown in *Figure 6.15*:

Figure 6.15 – Display showing the number 00789

In case of any problems, observe the serial port output for any errors. We recommend troubleshooting in the following sequence:

- Ensure that you have loaded up all the files as discussed in this section.
- Ensure that you have wired up everything correctly, including the wireless pack and the display's pins.
- Ensure that the Pico is connected to the wireless network.

Next, we will discuss driving the display via a serial port.

Serial port example

In this section, we will discuss driving the display via a serial port. If you are not familiar with serial ports, we recommend reading through *Chapter 2*, *Serial Interfaces and Applications*. This method is useful when the Pico is interfaced via USB to a Raspberry Pi. In this example, the Pico is used to just drive the display. The code sample discussed in this section is available for download along with this chapter's materials as `code_serial.py`:

- We make use of the seven-segment display drivers discussed earlier in this chapter but add a `while` loop to capture user input via a serial port:

```
while True:
    input_string = input("Enter a value to update?")
    if input_string[0] == "S":
        value_string = input_string[1:]
        try:
            value = int(value_string)
        except:
            print("Enter a valid number")
        else:
            show_number(value)
```

- We capture user input using the `input()` function. If the user input is of a particular format, for example, if the first character of the input is S, we extract the rest of the string and try converting it into an integer.

- If the conversion attempt fails, we ask for another input. Else, we update the display with the provided number.

- In this example, we are making use of the `input` function call, which blocks the program execution until the user input is available. If you want to avoid such blocking function calls, we recommend checking out the `supervisor` module (link: `https://bit.ly/3ozTgvT`) in CircuitPython. It enables calling the `input` function only when there is data available. We have provided an example that makes use of the `supervisor` module in this chapter's repository as `code_supervisor.py`.

We chose to discuss the serial port example because the display we originally built for Maker Faires made use of an Arduino and a Raspberry Pi. The Raspberry Pi fetched data from the cloud while the Arduino was used to drive the display. We wanted to demonstrate driving the display via a serial port in such applications.

Tracking physical activity

In order to display your daily step count from your fitness tracker, we need the Pico to connect to the cloud making use of the API provided by the manufacturer. Such APIs usually come with an authentication mechanism to access data. For example, Fitbit's fitness trackers come with an API that makes use of the **OAuth 2.0** mechanism (link to the Fitbit API's documentation: `https://bit.ly/3mmkHX3`). We are currently investigating making use of Fitbit's API to retrieve our physical activity data as well as handling refresh tokens. We will post our findings to this chapter's GitHub repository.

In the next section, we will discuss assembling the display in a shadow box.

Putting it all together

When you are done testing the display, it is time to assemble it. We initially planned the layout of the entire display on a sheet of plywood, as shown in *Figure 6.16*:

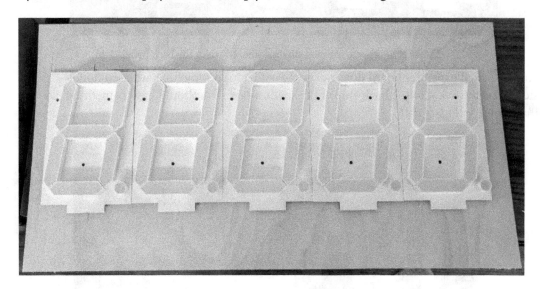

Figure 6.16 – Planning the layout of the seven-segment display

This enabled us to determine the dimensions of the shadow box needed for the project. We purchased a shadow box and assembled the digits on the back panel of the shadow box, as shown in *Figure 6.17*:

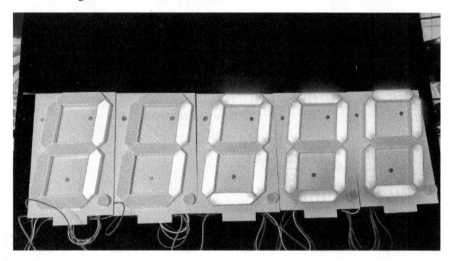

Figure 6.17 – Seven-segment digits assembled onto the back panel

Since the dual expander board does not come with a mounting hole, we used double-sided tape to stick it to the back side of the shadow box. You should have something like that shown in *Figure 6.18*:

Figure 6.18 – Giant seven-segment display

The shadow box made it easy to transport for Maker Faire exhibits. We usually power it using a 12 V DC adapter. When an internet connection is needed for the display to fetch data, we usually create a Wi-Fi hotspot that mimics the credentials we saved in `settings.py`.

Summary

In this chapter, we discussed building a giant seven-segment display. We discussed the drivers meant for driving the display. We also discussed planning a layout and assembling the display into a shadow box. This was followed by a review of how to use the display by connecting it to the internet and displaying numbers.

In the next chapter, we will discuss building an exhibit for tracking air quality.

7
Designing a Visual Aid for Tracking Air Quality

In the previous chapter, we built a visual aid containing a giant seven-segment display to track physical activity. In this chapter, we are going to build a visual aid that can serve as an interactive exhibit to track air quality. By the end of this chapter, you will be able to build something like the one shown in the following figure.

Figure 7.1 – Air quality exhibit

The goal of this project is to educate people about air quality in our immediate surroundings. We will discuss the project in two different ways. In the first approach, we will build the display using publicly available data sources. In the second approach, we will discuss building it using a carbon dioxide sensor that is interfaced with the **Raspberry Pi Pico**.

The topics discussed in this chapter include the following:

- Inspiration for the project
- Installing requisite libraries
- Using public data sources for air quality data
- Interfacing a CO_2 sensor with the Pico
- Interfacing the stepper motor
- Building the display

Technical requirements

The following hardware is recommended for this chapter:

- Raspberry Pi Pico (link: `https://bit.ly/3AJtoAf`) – USD 4.
- Pico Omnibus – Dual Expander (link: `https://bit.ly/3sr2GJR`) – USD 10.50.
- Pico Wireless Pack (link: `https://bit.ly/3yPuoT9`) – USD 16.75.
- Automotive Gauge Stepper Motor (link: `https://bit.ly/3FvHlFo`) – USD 9.95.
- Adafruit DRV8833 Stepper Motor Breakout (link: `https://bit.ly/3DkcgTb`) – USD 4.95.
- Jumper Wires (link: `https://amzn.to/3nmRI7L`) – USD 5.99.
- Adafruit SCD-30 CO2 sensor (optional) (link: `https://bit.ly/3l7MfQ6`) – USD 58.95.
- A shadow box from your local hobby store.
- The code samples for this chapter are available at `https://github.com/PacktPublishing/Raspberry-Pi-Pico-DIY-Workshop/tree/main/chapter_07`.

Code in Action videos for this chapter can be viewed at `https://bit.ly/3OXiHC5`.

> **CircuitPython Installation**
>
> We are assuming that you have installed CircuitPython on your Pico. If you are not familiar with the installation process, we recommend following the installation process from *Chapter 1, Getting Started with the Raspberry Pi Pico.*

Inspiration for the project

Good quality air is a prime right for humans everywhere. Increasing urbanization and global pollution have continued to lower the quality of air, which severely affects the quality of life for citizens everywhere, be it Peshawar, New Delhi, or anywhere else. Citizen scientists want to engage in monitoring air quality to help develop a better understanding of pollution levels and the elements that contribute to poor air quality. Parameters of interest include **Oxygen** **(O_2)** and **Carbon dioxide (CO_2)** levels, the presence of **Volatile Organic Compounds (VOCs)**, **Particulate Matter (PM)** arising from the residue of burnt coal, wood, and other materials, as well as compounds of **Sulphur**, **Nitrogen**, and other elements.

Sensors that monitor air quality are increasingly becoming lower in cost, while simultaneously continuing to gain in accuracy and precision. This will allow citizen scientists to develop solutions that allow the monitoring of air quality. The data collected from such efforts helps people understand, avert, reduce, and respond to pollution, thus saving and/or improving lives.

Installing requisite libraries

In this " section, we will install the requisite libraries needed for this chapter, including the stepper motor, the CO_2 sensor, and the wireless pack to the Raspberry Pi Pico. The libraries are all a part of the **Adafruit CircuitPython** bundle. The latest bundle can be downloaded as a ZIP file from `https://circuitpython.org/libraries`. We used the bundle version meant for CircuitPython 6.x.x.

After downloading the ZIP file, extract its contents so that we can copy the libraries we need for the project.

Stepper motor

We need the adafruit_motor library to control the stepper motor (shown in the following figure). The stepper motor has a needle that is used to point at the corresponding air quality category. Copy the adafruit_motor library to the lib folder of your Pico.

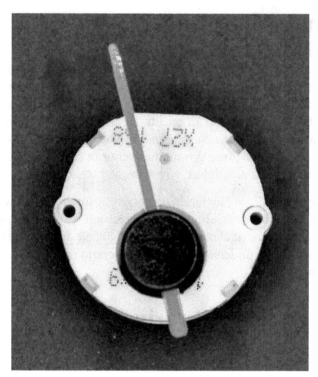

Figure 7.2 – Automotive gauge stepper motor

Next, we will review installing libraries for the wireless pack.

Wireless pack

We need the adafruit_esp32spi library for the wireless pack. Copy over the folder (with the same name) to the lib folder of your Pico. We will also need the adafruit_requests.mpy binary from the bundle.

> **Setting Up the Wireless Pack**
>
> In *Chapter 2, Serial Interfaces and Applications*, we discussed setting up the wireless pack in detail. We recommend reading through it and setting up your wireless pack.

SCD30 CO_2 sensor

This is an optional step. The CO_2 sensor (shown in the following figure) has the following dependencies: adafruit_bus_device, adafruit_register, and adafruit_scd30. Copy over the adafruit_bus_device and adafruit_register folders along with adafruit_scd30.mpy to the lib folder of your Pico.

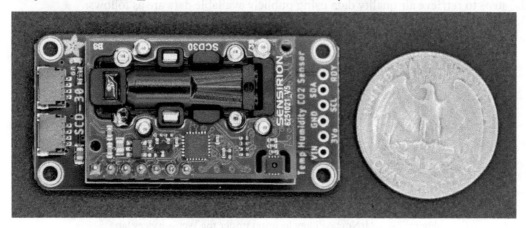

Figure 7.3 – SCD30 CO2 sensor

Now that we have installed the requisite libraries, we are going to discuss retrieving air quality data from public data sources.

Using public data sources for air quality data

In this section, we are going to discuss retrieving air quality data using publicly available data, namely the **AirNow API** provided by the **United States Environmental Protection Agency** (**US EPA**). API is the acronym for **Application Programming Interface**. Here, it refers to the interface provided to retrieve air quality data. An API typically comes with specifications for access. Here, it is a program that aggregates data from federal, state, and local agencies to report the **Air Quality Index** (**AQI**). We are going to test the API and then discuss the code needed to retrieve the air quality data using the Pico:

> **Data Sources Outside the United States**
>
> If you live outside the United States, we recommend using the local data source in your region. If one doesn't exist, we recommend checking out the *Interfacing a CO₂ sensor on a Pico* section of this chapter.

The steps to retrieve air quality data using publicly available data are as follows:

1. We recommend signing up for a free account at `https://airnowapi.org`. After signing into your account, you can find your API key in the top-right corner under the **Web Services** tab, as shown in the following screenshot:

Figure 7.4 – AirNow API key location under the Web Services tab

2. We are assuming that you have prepared your Pico for setting up the wireless pack. This includes setting up the Wi-Fi credentials in the file `secrets.py`. We need to include the AirNow API key in the following format:

```
secrets = {
    'ssid' : 'SSID',
    'password' : 'password',
    'timezone' : "America/New_York",
    'api_key'   : "ABCDEFGHIJK12345"
    }
```

3. We will be making use of the stored credentials to retrieve the local air quality data. Now let's revisit `https://airnowapi.org`. We are going to test the API using the tools available on the site before reviewing the code needed to retrieve air quality data using the Raspberry Pi Pico.

4. In your AirNow account, go to **Forecasts** on the **Web Services** tab, as shown in the following screenshot:

Figure 7.5 – Forecasts on the Web Services tab

5. We are going to make use of the query tool shown in the previous screenshot to create a URL that we are going to make use of in the code running on the Raspberry Pi Pico. We are going to select a zip code and the format in which we want the API to return the data. In this case, we are going to make use of the json format.

6. The following screenshot shows the query builder's parameters:

1	Zip Code:	14217	Distance:	25	miles
	Date:	2021-11-21	Format:	application/json	

Build

Figure 7.6 – Query builder's parameters

7. When you click on **Build**, it should generate a URL, as shown in the following screenshot:

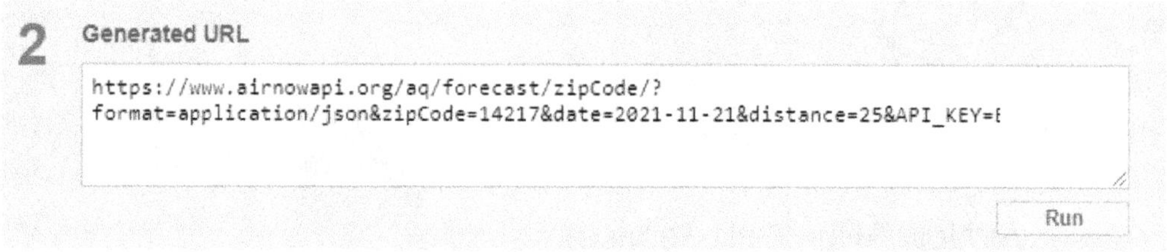

2   Generated URL

```
https://www.airnowapi.org/aq/forecast/zipCode/?
format=application/json&zipCode=14217&date=2021-11-21&distance=25&API_KEY=f
```

Run

Figure 7.7 – Building the query

8. The generated URL should look something like this:

```
https://www.airnowapi.org/aq/forecast/
zipCode/?format=application/json&zipCode=14217&date=2021-
11-21&distance=25&API_KEY=ABCEDFGHIJKL12356
```

9. Copy over the URL as we are going to need it for our code. Now, click on **Run** and you should see the air quality data for your zip code.

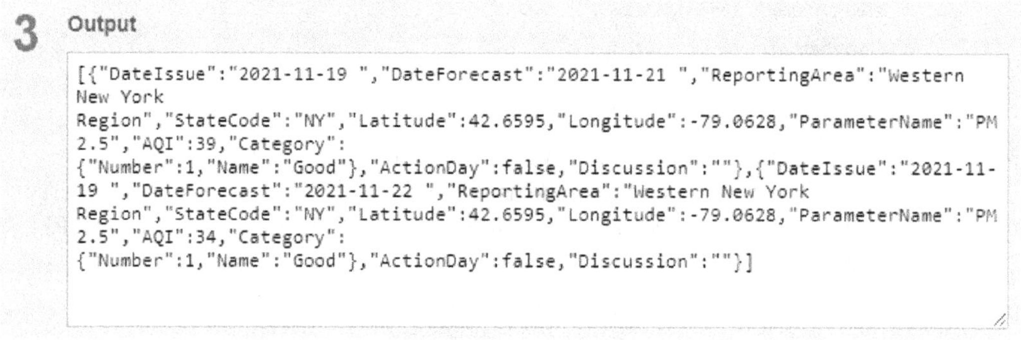

3   Output

```
[{"DateIssue":"2021-11-19 ","DateForecast":"2021-11-21 ","ReportingArea":"Western
New York
Region","StateCode":"NY","Latitude":42.6595,"Longitude":-79.0628,"ParameterName":"PM
2.5","AQI":39,"Category":
{"Number":1,"Name":"Good"},"ActionDay":false,"Discussion":""},{"DateIssue":"2021-11-
19 ","DateForecast":"2021-11-22 ","ReportingArea":"Western New York
Region","StateCode":"NY","Latitude":42.6595,"Longitude":-79.0628,"ParameterName":"PM
2.5","AQI":34,"Category":
{"Number":1,"Name":"Good"},"ActionDay":false,"Discussion":""}]
```

Figure 7.8 – Air quality data

10. Let's take a closer look at the response:

```
{"DateIssue":"2021-11-19 ","DateForecast":"2021-11-21
","ReportingArea":"Western New York Region","StateCode":"
NY","Latitude":42.6595,"Longitude":-79.0628,"ParameterNam
e":"PM2.5","AQI":39,"Category":{"Number":1,"Name":"Good"}
,"ActionDay":false,"Discussion":""}
```

In the preceding response snippet, we can see that the AQI is 39 (highlighted in the response) and the category is Good. The category is referring to one of the six categories of air quality (shown in the following screenshot). The categories help understand how the measured air quality relates to public health.

AQI Numbers	AQI Category (Descriptor)	AQI Color	Hexadecimal Color Value	Category Number
0 - 50	Good	Green	(00e400)	1
51 - 100	Moderate	Yellow	(ffff00)	2
101 - 150	Unhealthy for Sensitive Groups	Orange	(ff7e00)	3
151 - 200	Unhealthy	Red	(ff0000)	4
201 - 300	Very Unhealthy	Purple	(8f3f97)	5
301 - 500	Hazardous	Maroon	(7e0023)	6

Figure 7.9 – Air Quality categories. Source: airnowapi.org

In the query response, the air quality is **Good**. When we run the same query on a Pico, we will retrieve the air quality category and use it to drive a stepper motor to point to the corresponding category. This visual aid can be used to educate people about the air quality in their community.

We recommend trying different zip codes across continental United States to check out the air quality in various locations. The US EPA provides specific recommendations in areas prone to wildfires.

Now, let's discuss running the query on a Pico.

Running the query on a Pico

In this section, we are going to discuss the code needed to run the query on a Pico using the **ESP32 wireless pack**. Ensure that you have installed the requisite packages discussed earlier in this chapter. In this example, we are going to make the code execution interactive, where we take user inputs for local zip codes and dates. The code is available for download along with this chapter as code_airnow.py:

1. The first step is to import the requisite modules needed:

```
import board
import busio
from digitalio import DigitalInOut
```

```
import adafruit_requests as requests
import adafruit_esp32spi.adafruit_esp32spi_socket as
socket
from adafruit_esp32spi import adafruit_esp32spi
```

2. The next step is to import the Wi-Fi credentials from `secrets.py`:

```
try:
    from secrets import secrets
except ImportError:
    print("WiFi secrets are kept in secrets.py, please
add them there!")
    raise
```

3. Next, we declare the `API_URL` variable without the zip code, date, or *API key*. Copy over the URL from the previous section and save it in the following format:

```
API_URL = "https://www.airnowapi.org/aq/
forecast/zipCode/?format=application/
json&zipCode={0}&date={1}&distance=25&API_KEY={2}"
```

4. We will also load the *API key* stored in `secrets.py` as follows:

```
API_KEY = secrets["api_key"]
```

5. The next step is to initialize the pins needed for the wireless pack and initialize the **serial peripheral interface** (**SPI**):

```
esp32_cs = DigitalInOut(board.GP7)
esp32_ready = DigitalInOut(board.GP10)
esp32_reset = DigitalInOut(board.GP11)
spi = busio.SPI(board.GP18, board.GP19, board.GP16)
```

6. Then, we initialize the ESP32 wireless pack:

```
esp = adafruit_esp32spi.ESP_SPIcontrol(spi, esp32_cs,
esp32_ready, esp32_reset)
requests.set_socket(socket, esp)


if esp.status == adafruit_esp32spi.WL_IDLE_STATUS:
    print("ESP32 found and in idle mode")
```

```
print("Firmware vers.", esp.firmware_version)
print("MAC addr:", [hex(i) for I in esp.MAC_address])
```

7. Now, we connect to the wireless network using the credentials saved in `secrets.py`:

```
print("Connecting to AP...")
while not esp.is_connected:
    try:
        esp.connect_AP(secrets["ssid"],
secrets["password"])
    except RuntimeError as e:
        print("could not connect to AP, retrying: ", e)
        continue

print("Connected to", str(esp.ssid, "utf-8"), "\tRSSI:",
esp.rssi)
print("My IP address is", esp.pretty_ip(esp.ip_address))
```

8. Once the wireless pack is connected to the internet, we need to capture the user inputs for the zip code and date:

```
zipcode = input("Enter a valid 5-digit zipcode: ")
date = input("Enter today's date in the following format
YYYY-MM-DD: ")
```

9. With the captured inputs, we can format our `API_URL` using `API_KEY`:

```
full_url = API_URL.format(zipcode, date, API_KEY)
```

10. Now, we can retrieve the local air quality data using a `try-except-else` block. If the GET request was successful, the code under `else` is executed. If there were any errors, the exception is captured by the `except` block:

```
try:
    response = requests.get(full_url)
except Exception as e:
    print(e)
else:
    data = response.json()[0]
    print("-" * 40)
```

```
        print("Reporting Area: ", data["ReportingArea"])
        print("AQI: ", data["AQI"])
        print("Category: ", data["Category"]["Name"])
        print("-" * 40)
        response.close()
```

11. In the `else` block, we retrieve the data and extract the following information: `Reporting Area`, `AQI`, and the `Category` of air quality from the response.

12. Putting it all together, we have the following (also available for download as `code_airnow.py`):

```python
import board
import busio
from digitalio import DigitalInOut
import adafruit_requests as requests
import adafruit_esp32spi.adafruit_esp32spi_socket as socket
from adafruit_esp32spi import adafruit_esp32spi


# Get Wi-Fi details and more from a secrets.py file
try:
    from secrets import secrets
except ImportError:
    print("WiFi secrets are kept in secrets.py, please add them there!")
    raise


API_URL = "https://www.airnowapi.org/aq/forecast/zipCode/?format=application/json&zipCode={0}&date={1}&distance=25&API_KEY={2}"
API_KEY = secrets["api_key"]


# If you are using a board with pre-defined ESP32 Pins:
esp32_cs = DigitalInOut(board.GP7)
esp32_ready = DigitalInOut(board.GP10)
esp32_reset = DigitalInOut(board.GP11)


spi = busio.SPI(board.GP18, board.GP19, board.GP16)
```

```python
esp = adafruit_esp32spi.ESP_SPIcontrol(spi, esp32_cs,
esp32_ready, esp32_reset)
requests.set_socket(socket, esp)


if esp.status == adafruit_esp32spi.WL_IDLE_STATUS:
    print("ESP32 found and in idle mode")


print("Connecting to AP...")
while not esp.is_connected:
    try:
        esp.connect_AP(secrets["ssid"],
secrets["password"])
    except RuntimeError as e:
        print("could not connect to AP, retrying: ", e)
        continue


print("Connected to", str(esp.ssid, "utf-8"), "\tRSSI:",
esp.rssi)
print("My IP address is", esp.pretty_ip(esp.ip_address))


zipcode = input("Enter a valid 5-digit zipcode: ")
date = input("Enter today's date in the following format
YYYY-MM-DD: ")


full_url = API_URL.format(zipcode, date, API_KEY)


try:
    response = requests.get(full_url)
except Exception as e:
    print(e)
else:
    data = response.json()[0]
    print("-" * 40)
    print("Reporting Area: ", data["ReportingArea"])
    print("AQI: ", data["AQI"])
    print("Category: ", data["Category"]["Name"])
    print("-" * 40)
```

```
      response.close()


  print("Done!")
```

13. When we save the preceding sample as `code.py` on the Pico and run it, we should be able to provide inputs to the program, namely the `zipcode` and the `date` for the forecast.

```
Enter a valid 5-digit zipcode: 94103
Enter today's date in the following format YYYY-MM-DD: 2021-10-23
```

Figure 7.10 – Providing user inputs to the AirNow API program

14. The `zipcode` provided as an input belongs to the city of San Francisco. We should see an `AQI` forecast as shown in the following screenshot:

```
----------------------------------------
Reporting Area:   San Francisco
AQI:   42
Category:   Good
----------------------------------------
```

Figure 7.11 – Air Quality Index forecast for San Francisco on Nov 23 2021

15. We ran the air quality checks for various zip codes. For example, the air quality forecast on the same day in Buffalo, NY was as follows:

```
----------------------------------------
Reporting Area:   Western New York Region
AQI:   29
Category:   Good
----------------------------------------
```

Figure 7.12 – Air Quality Index forecast for Buffalo, NY on Nov 23 2021

In the previous two screenshots, you can notice the substantial difference in the indices between the two cities despite being in the Good category. This tool can be used to educate people on air quality in their community.

We originally built this project as a Maker Faire exhibit where we made use of a keypad to let the participant enter their zip code to learn about the air quality in the community. We also demonstrated by showing zip codes where the air quality forecasts came with warnings for the general public. Weather forecasters use this information to warn people with allergies in the spring season.

In this example, the code executes only once per user input. This is because we wanted the program execution to be interactive. If this project is going to be used in a permanent installation, we will hardcode the zip code and fetch the air quality at regular intervals. We also need to fetch the local network date and time using the wireless pack. We have demonstrated this in the code sample `code_airnow_loop.py` (available for download along with this chapter).

The AirNow API calls to fetch air quality by zip code are rate-limited at `500` calls per hour. You need to be careful while making use of the API as it is a public service. Now that we are able to retrieve the air quality data, we can make use of the fetched data to build our exhibit. In the next section, we will work on interfacing a CO_2 sensor with the Pico.

Interfacing a CO_2 sensor with the Pico

In this section, we will discuss interfacing the CO_2 sensor with the Raspberry Pi Pico. This section is meant for those who don't have access to a local data source for air quality. The SCD30 sensor has a measurement range of 400–10,000 ppm. We chose this sensor because it comes with an onboard temperature sensor that provides temperature compensation to the CO_2 concentration calculation. The datasheet for the sensor is available from here: `https://bit.ly/3tyJZ9C`. We will measure the CO_2 concentration and publish it to **ThingSpeak**, a service we discussed in *Chapter 4, Fun with Gardening!*. This will enable us to share a public dashboard of the local air quality levels.

If you have access to a public data source, you are welcome to skip this section and move to the next section where we test the motors.

We are assuming that you have installed the required libraries to communicate with the sensor using the instructions from earlier in this chapter.

The carbon dioxide sensor comes with an I2C interface and is interfaced to the Pico as shown in *Figure 7.13*. The connections are listed here such that the left-hand side of the arrow refers to the pin on the Pico while the right refers to a pin on the CO_2 sensor:

- GP9 → SCL
- GP8 → SDA
- 3.3V → VIN
- GND pins tied together

Pull-Up Resistors

The SCD30 breakout board comes with the pull-up resistors needed for the I2C interface.

The following figure shows the Fritzing schematic for interfacing the CO_2 sensor to the Raspberry Pi Pico:

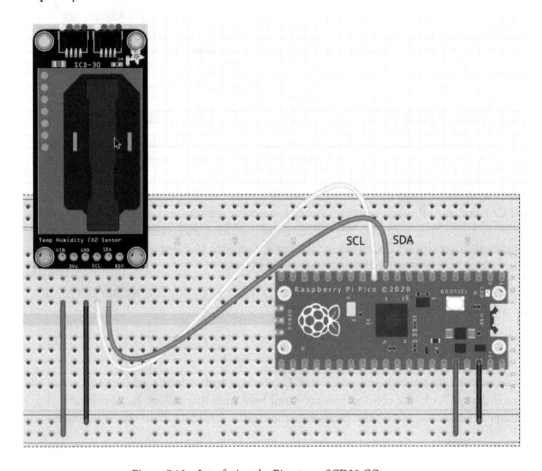

Figure 7.13 – Interfacing the Pico to an SCD30 CO_2 sensor

Now, let's review the code needed to interface the sensor. The code sample discussed here is available for download along with this chapter as `code_scd30.py`:

1. The first step is to import the requisite modules:

```
import board
import busio
import adafruit_scd30
```

2. Next, we will initialize the I2C interface followed by the CO_2 sensor:

```
i2c = busio.I2C(board.GP9, board.GP8)
scd = adafruit_scd30.SCD30(i2c)
```

3. In the following code snippet, you will note that the text is split across two lines. There are actually no line breaks but the text formatting splits it across two lines. Now, we will enter a `while` loop where we will print the CO_2 concentration if an update is available:

```
while True:
    if scd.data_available:
        print("Data Available!")
        print("CO2: %d PPM" % scd.CO2)
        print("Temperature: %0.2f degrees C" % scd.
temperature)
        print("Humidity: %0.2f %% rH" % scd.relative_
humidity)
```

4. Putting it all together, we have the following:

```
import time
import board
import busio
import adafruit_scd30


i2c = busio.I2C(board.GP9, board.GP8)
scd = adafruit_scd30.SCD30(i2c)


while True:
    if scd.data_available:
        print("Data Available!")
```

```
    print("CO2: %d PPM" % scd.CO2)
    print("Temperature: %0.2f degrees C" % scd.
temperature)
    print("Humidity: %0.2f %% rH" % scd.relative_
humidity)


    time.sleep(2.0)
```

5. When we save the preceding code sample as code.py on the Pico, we should be able to see the following output:

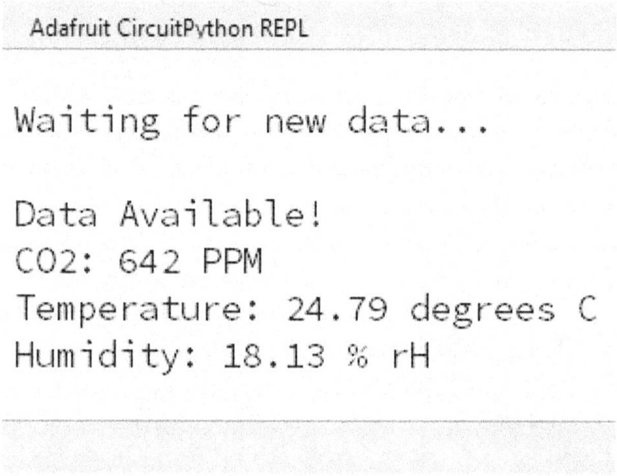

Figure 7.14 – CO$_2$ sensor output

Ease of Prototyping

If you have been following along with us, you will have noticed the ease of interfacing sensors and reading data using CircuitPython. It makes it easier to build our projects.

Now, we will discuss publishing the sensor data to **ThingSpeak** – a platform we used earlier, in *Chapter 4, Fun with Gardening!*. We recommend that you create a free account on ThingSpeak (if you haven't already).

Let's take a look at the steps involved in publishing data to ThingSpeak:

1. The first step is to create a new channel on ThingSpeak (as shown in the following screenshot). You can only create a maximum of four channels using a free account.

My Channels

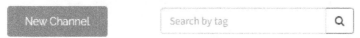

Figure 7.15 – Create a new channel on ThingSpeak

2. Create a field called CO2 concentration as shown in the following screenshot:

New Channel

Figure 7.16 – Create a new channel

3. Once the channel is created, copy the write API key and the URL from the **Write a Channel Feed**. This is available under the **API Keys** tab of your channel.

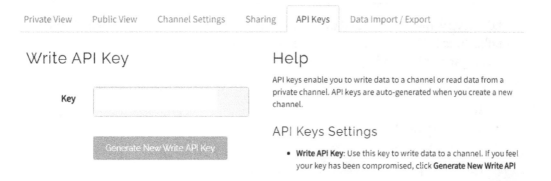

Figure 7.17 – Copy the write API key location

4. Also, make a note of the URL needed to publish the sensor data. It is available on the same tab.

Write a Channel Feed

```
GET https://api.thingspeak.com/update?api_key=
```

Figure 7.18 – Copy the write URL feed

5. Now, we will modify secrets.py to include the ThingSpeak API key:

```
secrets = {
    'ssid' : 'SSID',
    'password' : 'password',
    'timezone' : "America/New_York",
    'api_key'  : "ABCDEFGHIJK12345"
    }
```

> **API Key**
>
> We are assuming that you are following this example due to the lack of local data sources to pursue the previous example. We are storing the API key under the field api_key.

6. Let's review the code needed to publish data to ThingSpeak. The code sample for publishing to ThingSpeak is available for download along with this chapter as code_scd_thingspeak.py:

```
import board
import busio
import time
import adafruit_scd30
from digitalio import DigitalInOut
import adafruit_requests as requests
import adafruit_esp32spi.adafruit_esp32spi_socket as socket
from adafruit_esp32spi import adafruit_esp32spi
```

7. The next step is to import the Wi-Fi credentials from `secrets.py`:

```
try:
    from secrets import secrets
except ImportError:
    print("WiFi secrets are kept in secrets.py, please
add them there!")
    raise
```

8. Next, we declare the `API_URL` variable without the zip code, date, or *API key*. Copy over the URL from the previous section and save it in the following format:

```
API_URL = "https://api.thingspeak.com/update?api_
key={0}&field1={1}"
```

9. We will also load the *API key* stored in `secrets.py` as follows:

```
API_KEY = secrets["api_key"]
```

10. The next step is to initialize the pins needed for the wireless pack and initialize the SPI:

```
esp32_cs = DigitalInOut(board.GP7)
esp32_ready = DigitalInOut(board.GP10)
esp32_reset = DigitalInOut(board.GP11)
spi = busio.SPI(board.GP18, board.GP19, board.GP16)
```

11. Then, we initialize the ESP32 wireless pack:

```
esp = adafruit_esp32spi.ESP_SPIcontrol(spi, esp32_cs,
esp32_ready, esp32_reset)
requests.set_socket(socket, esp)
```

12. Then we initialize the I2C interface and the SCD30 CO_2 sensor:

```
i2c = busio.I2C(board.GP9, board.GP8)
scd = adafruit_scd30.SCD30(i2c)
```

13. Next, we enter a `while` loop where we connect to the wireless network if we haven't already:

```
while True:
    while not esp.is_connected:
        try:
```

```
        print("Connecting to AP...")
        esp.connect_AP(secrets["ssid"],
secrets["password"])
    except RuntimeError as e:
        print("could not connect to AP, retrying: ",
e)
        continue
    else:
        print("Connected to", str(esp.ssid, "utf-8"),
"\tRSSI:", esp.rssi)
        print("My IP address is", esp.pretty_ip(esp.
ip_address))
```

14. Then, when the CO_2 is ready, we read the CO_2 concentration:

```
    if scd.data_available:
        print("Data Available!")
        print("CO2: %d PPM" % scd.CO2)
        print("Temperature: %0.2f degrees C" % scd.
temperature)
        print("Humidity: %0.2f %% rH" % scd.relative_
humidity)
```

15. Then, we publish the data to ThingSpeak:

```
        full_url = API_URL.format(API_KEY, str(scd.CO2))
        try:
            response = requests.get(full_url)
        except Exception as e:
            print(e)
        else:
            print("-" * 40)
            print(response.json())
            print("-" * 40)
            response.close()
            print("Done!")
    time.sleep(20)
```

16. Putting it altogether, we have the following:

```
import board
import busio
import time
import adafruit_scd30
from digitalio import DigitalInOut
import adafruit_requests as requests
import adafruit_esp32spi.adafruit_esp32spi_socket as
socket
from adafruit_esp32spi import adafruit_esp32spi


# Get wifi details and more from a secrets.py file
try:
    from secrets import secrets
except ImportError:
    print("WiFi secrets are kept in secrets.py, please
add them there!")
    raise

API_URL = "https://api.thingspeak.com/update?api_
key={0}&field1={1}"
API_KEY = secrets["api_key"]

# If you are using a board with pre-defined ESP32 Pins:
esp32_cs = DigitalInOut(board.GP7)
esp32_ready = DigitalInOut(board.GP10)
esp32_reset = DigitalInOut(board.GP11)

spi = busio.SPI(board.GP18, board.GP19, board.GP16)
esp = adafruit_esp32spi.ESP_SPIcontrol(spi, esp32_cs,
esp32_ready, esp32_reset)
requests.set_socket(socket, esp)

i2c = busio.I2C(board.GP9, board.GP8)
scd = adafruit_scd30.SCD30(i2c)
```

```python
while True:
    while not esp.is_connected:
        try:
            print("Connecting to AP...")
            esp.connect_AP(secrets["ssid"],
secrets["password"])
        except RuntimeError as e:
            print("could not connect to AP, retrying: ",
e)
            continue
        else:
            print("Connected to", str(esp.ssid, "utf-8"),
"\tRSSI:", esp.rssi)
            print("My IP address is", esp.pretty_ip(esp.
ip_address))

    if scd.data_available:
        print("Data Available!")
        print("CO2: %d PPM" % scd.CO2)
        print("Temperature: %0.2f degrees C" % scd.
temperature)
        print("Humidity: %0.2f %% rH" % scd.relative_
humidity)
        full_url = API_URL.format(API_KEY, str(scd.CO2))

        try:
            response = requests.get(full_url)
        except Exception as e:
            print(e)
        else:
            print("-" * 40)
            print(response.json())
            print("-" * 40)
            response.close()
            print("Done!")
    time.sleep(20)
```

17. When we save the preceding sample as code.py on the Pico and run it, we should see the CO$_2$ concentration and an acknowledgement of the data being published to ThingSpeak. We should see an output similar to the one shown in the following screenshot:

```
Data Available!
CO2: 634 PPM
Temperature: 27.86 degrees C
Humidity: 23.46 % rH
-----------------------------
339
-----------------------------
Done!
```

Figure 7.19 – Temperature data published to ThingSpeak

18. We should also be able to see CO$_2$ concentration trends over time as shown in the following screenshot:

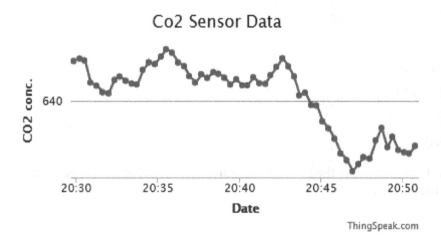

Figure 7.20 – Publishing CO$_2$ data to ThingSpeak

We should be able to share this data as a dashboard to educate everyone about the CO$_2$ concentration in their immediate vicinity. Likewise, we can add a VOC sensor, PM sensor, and have a dashboard that tracks all these monitors on a giant screen.

In the next section, we will discuss controlling the stepper motor for our exhibit.

Interfacing the stepper motor

In this section, we will discuss interfacing a stepper motor with the Raspberry Pi Pico using the **DRV8833 Stepper Motor Driver**. This will enable us to build a display where the stepper motor will point to the corresponding air quality category.

> **Motor Selection**
>
> For this project, we chose a stepper motor that is used in automotive gauges. This is because the motor has a total movement range of 315°. It also comes with a needle that is useful in building our visual aid. Motor selection for an application involves several factors. We recommend the following article from SparkFun: `https://bit.ly/3qyYxnX`.

The first step is to wire up the stepper motor driver to the Raspberry Pi Pico. *Figure 7.21* shows the connections of the stepper motor driver to the Pico. The stepper motor driver is connected as follows where the left-hand side of the arrow refers to a pin available on the Raspberry Pi Pico while the right refers to a pin on the DRV8833 stepper motor driver:

- 3.3V → VM, SLP
- GP12 → AIN1
- GP13 → AIN2
- GP14 → BIN1
- GP15 → BIN2
- GND

The following figure shows the Fritzing schematic for interfacing the stepper motor driver with the Raspberry Pi Pico:

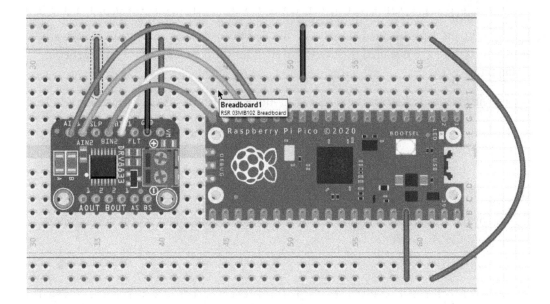

Figure 7.21 – Interface DRV8833 with the Raspberry Pi Pico

Now, the stepper motor comes with two pairs of coils. The first pair is connected to **AOUT** while the second pair is connected to **BOUT**. In case you are not familiar with stepper motors, we recommend checking out this article: `https://bit.ly/3A6mfuN`. The connections to the stepper motor on a **breadboard** are shown in the following figure:

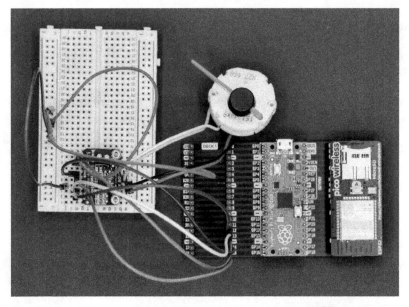

Figure 7.22 – Interfacing the stepper motor to the DRV8833 driver

Now, let's look at the code needed to drive a stepper motor. The code discussed in this section is available for download along with this chapter as `code_drv8833.py`. We are assuming that you have installed the drivers needed for the stepper motor from earlier in this chapter.

1. The first step is to import the requisite modules for driving the stepper motor:

```
import time
import board
import digitalio
from adafruit_motor import stepper
```

2. The next step is to declare the pins used for driving the stepper motor and set them as output pins:

```
coils = (
    digitalio.DigitalInOut(board.GP12),  # A1
    digitalio.DigitalInOut(board.GP13),  # A2
    digitalio.DigitalInOut(board.GP14),  # B1
    digitalio.DigitalInOut(board.GP15),  # B2
)
for coil in coils:
    coil.direction = digitalio.Direction.OUTPUT
```

3. We initialize the `StepperMotor` class from the stepper motor driver libraries:

```
motor = stepper.StepperMotor(coils[0], coils[1],
coils[2], coils[3], microsteps=None)
```

4. Now, we drive the stepper motor forwards and backwards for 400 steps each way with a 10 ms delay between each step:

```
DELAY = 0.01
STEPS = 400
for step in range(STEPS):
    motor.onestep()
    time.sleep(DELAY)

for step in range(STEPS):
    motor.onestep(direction=stepper.BACKWARD)
    time.sleep(DELAY)
```

5. At the end of the program, we finally release the GPIO pins used to drive the stepper motor by calling `motor.release()`.

6. Putting it all together, we have the following:

```python
import time
import board
import digitalio
from adafruit_motor import stepper


DELAY = 0.01
STEPS = 400


coils = (
    digitalio.DigitalInOut(board.GP12),   # A1
    digitalio.DigitalInOut(board.GP13),   # A2
    digitalio.DigitalInOut(board.GP14),   # B1
    digitalio.DigitalInOut(board.GP15),   # B2
)


for coil in coils:
    coil.direction = digitalio.Direction.OUTPUT


motor = stepper.StepperMotor(coils[0], coils[1],
coils[2], coils[3], microsteps=None)


for step in range(STEPS):
    motor.onestep()
    time.sleep(DELAY)


for step in range(STEPS):
    motor.onestep(direction=stepper.BACKWARD)
    time.sleep(DELAY)


motor.release()
```

If the connections between the stepper motor driver and the Pico as well as the stepper motor are correct, we should see the needle atop the stepper motor move back and forth. In the next section, we will discuss assembling everything so that we can use the needle to point at the air quality category.

Building the display

It is time to assemble the display inside the shadow box. We printed the various air quality categories on a semi-circle and tested its fit inside the shadow box as shown in the following figure. If you are not familiar with shadow boxes, they are display boxes with a recess and glass at the front. They are used for prominently displaying souvenirs, medals, mementos, and so on. They are usually carried by stores that sell supplies for arts, crafts, and other related hobbies.

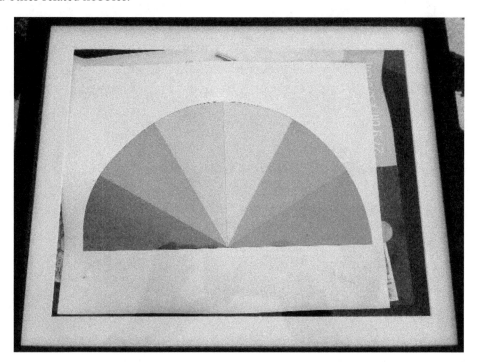

Figure 7.23 – Testing the fit inside the shadow box

We ensure that the stepper motor sits in the middle of the shadow box on top of the air quality categories. Once the stepper motor is installed, it is time to calibrate the stepper motor movements and calculate the number of steps needed for each category.

Now we are ready to take the display for a spin. The following figure shows the display pointing to the air quality being **Good**:

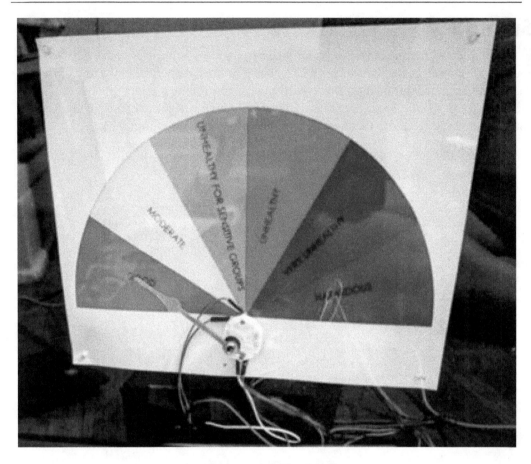

Figure 7.24 – Completed display

We are almost there with our project. In the next section, we are going to discuss making the display interactive.

Making an interactive display

If you are planning to take it to a *Maker Faire*, you can make it interactive by adding a keypad (link: `https://www.sparkfun.com/products/15290`) to the display. You can ask visitors to enter their zip code and learn more about the air quality in their community. You can also pick up some air quality forecasts on the day of the Faire and talk about how it impacts the people living in that community. For example, when the air quality is bad, the local authorities could place temporary restrictions on outdoor activities. It could be a great tool to get people motivated with electronics and citizen science.

Summary

In this chapter, we discussed retrieving air quality data from publicly available data sources. We also discussed interfacing a CO_2 sensor to the Raspberry Pi Pico to build a dashboard that tracks air quality. Then, we discussed controlling a stepper motor using the Pico, which can be used to point at one among six different categories of air quality. This enables us to build a display that keeps track of air quality in our community.

In the next chapter, we are going to have fun with radios! Join us in the adventure.

Section 3: Advanced Topics

In this section, we will embark on discussing advanced topics using the Raspberry Pi Pico. With experience gained from the previous sections, we are going to discuss topics ranging from interfacing wireless modules to building your own product using the Pico.

This section contains the following chapters:

- *Chapter 8, Building Wireless Nodes*
- *Chapter 9, Let's Build a Robot!*
- *Chapter 10, Designing TinyML Applications*
- *Chapter 11, Let's Build a Product!*
- *Chapter 12, Best Practices for Working with the Pico*

8

Building Wireless Nodes

In this chapter, we are going to discuss building wireless nodes using a Raspberry Pi Pico. So far, we have discussed publishing sensor data using a Wi-Fi module. What if we want to install a project at a location where we don't have access to a Wi-Fi network? In this chapter, we will discuss alternatives for such a scenario.

In this chapter, we are going to cover the following topics:

- Installing the requisite libraries
- Interfacing a Bluetooth Low Energy module
- Interfacing a Sigfox module
- Interfacing a LoRa module

> **Cellular Modules**
>
> We will discuss interfacing cellular modules in *Chapter 11, Let's Build a Product!*, where we will discuss building a product using a Raspberry Pi Pico and interfacing a cellular module.

Technical requirements

The following hardware is recommended for this chapter. We recommend purchasing components according to the type of wireless radio you want to use in your project:

- Raspberry Pi Pico (`https://bit.ly/3AJtoAf`) – 4 USD

- Optional – Adafruit Bluefruit LE SPI Friend (`https://bit.ly/3cXjukF`) – 17.50 USD

- Optional – Adafruit RFM95W Transceiver Breakout (`https://bit.ly/3y2UDWo`) –19.95 USD

- Jumper wires (`https://amzn.to/3nmRI7L`) – 5.99 USD

- Optional – Adafruit SCD-30 CO2 sensor (`https://bit.ly/317MfQ6`) – 58.95 USD

- A breadboard from your local electronics supplies store

The code samples for this chapter are available here: `https://github.com/PacktPublishing/Raspberry-Pi-Pico-DIY-Workshop/tree/main/chapter_08`.

Code in Action videos for this chapter can be viewed at `https://bit.ly/3Fgslvi`.

> **CircuitPython Installation**
>
> We are assuming that you have installed CircuitPython on your Pico. If you are not familiar with the installation process, we recommend following the installation process from *Chapter 1*, *Getting Started with the Raspberry Pi Pico*.
>
> **Optional components**: We have written this chapter in a choose-your-own-adventure style. You only need the components for the type of wireless radio project you are planning to pursue.

In the next section, we will discuss installing the requisite libraries needed for the three wireless radio examples in this chapter.

Installing requisite libraries

In this section, we will install the requisite libraries needed for the examples discussed in this chapter involving the Bluetooth, LoRa, and the Sigfox modules. The libraries are all part of the Adafruit CircuitPython bundle. The latest bundle can be downloaded as a ZIP file from `https://circuitpython.org/libraries`. We used the bundle version meant for CircuitPython 6.x.x.

After downloading the ZIP file, extract the contents so that we can copy the libraries we need for the project.

Adafruit Bluefruit LE SPI Friend

We will refer to the Adafruit Bluefruit LE SPI Friend (shown in the following photo) as the Bluetooth module for the rest of this chapter:

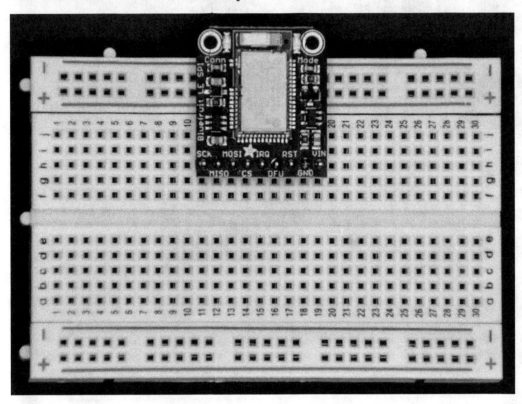

Figure 8.1 – Adafruit Bluefruit LE SPI Friend

The Bluetooth module has the `adafruit_bus_device` and `adafruit_bluefruitspi` dependencies. From the library bundle, copy over the `adafruit_bus_device` folder and the `adafruit_bluefruitspi.mpy` binary to the `lib` folder of the Pico.

Optional – the LoRa module

This is an optional step. You need to install the libraries associated with the LoRa module (shown in the following photo) only if you end up using it:

Figure 8.2 – LoRa module

The LoRa module needs the `adafruit_bus_device` and `adafruit_rfm9x` libraries. Copy over the `adafruit_bus_device` folder along with the `adafruit_rfm9x` binary to the `lib` folder of your Pico.

Optional – the CO2 sensor

This is an optional step. You need to install the CO2 sensor library only if you end up using the sensor. The CO2 sensor (shown in the following photo) has the `adafruit_bus_device`, `adafruit_register`, and `adafruit_scd30` dependencies. Copy over the `adafruit_bus_device` and `adafruit_register` folders along with `adafruit_scd30.mpy` to the lib folder of your Pico:

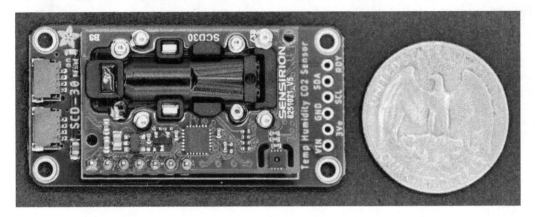

Figure 8.3 – The SCD30 CO2 sensor

In the next section, we will discuss interfacing a Bluetooth module to the Raspberry Pi Pico.

Interfacing a Bluetooth Low Energy module

In this section, we will discuss interfacing the Adafruit Bluefruit LE SPI Friend (hereby referred to as the Bluetooth module) to the Raspberry Pi Pico.

In case you are not familiar with Bluetooth Low Energy technology, it is a low-power wireless network technology meant for wireless sensors and other peripherals that can run off a coin cell. You can learn more about **Bluetooth Low Energy** (**BLE**) from this tutorial: `https://bit.ly/3BkLPNu`.

The breakout board used in this section comes with a Raytac MDBT40 BLE module, which is FCC and CE certified. In case you are not familiar with FCC and CE certification, they are mandatory certifications of approval for use in the United States and Europe respectively. The module is designed around the Nordic nRF51822 chipset. The folks at Adafruit implemented firmware that enables you to interface a Bluetooth module as a peripheral in your project. For more information, including the datasheet of the module, we recommend this article from Adafruit: `https://bit.ly/3sFaWX4`.

We will demonstrate applications where we can use the Bluetooth module as a peripheral to the Pico. We will be interfacing the Bluetooth module using the **Serial Peripheral Interface (SPI)** to the Pico. In the first example, we will transmit some test strings via Bluetooth to a mobile device. The transmitted message is echoed back to the module. It is then read by the Pico and printed to the screen. In the second example, we will discuss plotting sensor data on your mobile app. Let's get started!

The pin connects between the Pico and the Bluetooth module are as follows (the left side of the arrow refers to a pin on the Pico while the right refers to the Bluetooth module):

- VBUS → VIN
- GP0 → IRQ
- GP1 → RST
- GP5 → CS
- GP2 → SCK
- GP3 → MOSI
- GP4 → MISO
- GND pins tied together

The Fritzing schematic to interface the Bluetooth module with Raspberry Pi Pico is shown in the following figure:

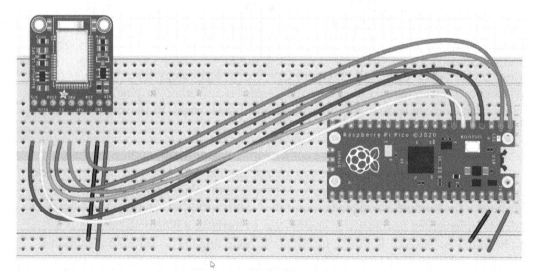

Figure 8.4 – Interfacing the Bluetooth module with the Raspberry Pi Pico

Now, let's discuss the code needed to interface the Bluetooth module. The example discussed in this section is based on the documentation available for the `adafruit_bluefruitspi` library. The code sample discussed in this section is available for download in the `downloads` folder of this chapter's GitHub repository as `bluefruit_spi_test.py`. The steps include the following:

1. The first step is to import the requisite modules for the program. They are as follows:

```
import time
import busio
import board
from digitalio import DigitalInOut
from adafruit_bluefruitspi import BluefruitSPI
```

2. The next step is to initialize the SPI interface and set up the Bluetooth module:

```
spi_bus = busio.SPI(board.GP2, MOSI=board.GP3,
MISO=board.GP4)
cs = DigitalInOut(board.GP5)
irq = DigitalInOut(board.GP0)
rst = DigitalInOut(board.GP1)
bluefruit = BluefruitSPI(spi_bus, cs, irq, rst,
debug=False)
```

3. Then, we initialize the Bluetooth module using an `AT` command. In case you are not familiar with `AT` commands, they are a command set used to control modems, where **AT** stands for **attention**. In this case, the folks at Adafruit developed firmware to control the Bluetooth module using `AT` commands:

```
print("Initializing the Bluefruit LE SPI Friend module")
bluefruit.init()
bluefruit.command_check_OK(b'AT+FACTORYRESET', delay=1)
```

4. We request information from the Bluetooth module:

```
print(str(bluefruit.command_check_OK(b'ATI'), 'utf-8'))
```

5. Bluetooth modules advertise with a name in order to connect to the device. Since we are going to look at the data on a mobile phone, we need to set a name that is easy to identify on a mobile phone application. Let's set this name as `PicoBLE`.

```
bluefruit.command_check_OK(b'AT+GAPDEVNAME=PicoBLE')
```

6. Now, we enter our main loop, where we wait for a device to connect to the Bluetooth module. Since we are awaiting a connection, we will be printing a " . " character every 500 ms to the serial terminal (as shown in *Figure 8.5*):

```
while True:
    print("Waiting for a connection to Bluefruit LE
Connect ...")
    dotcount = 0
    while not bluefruit.connected:
        print(".", end="")
        dotcount = (dotcount + 1) % 80
        if dotcount == 79:
            print("")
        time.sleep(0.5)
```

This helps the user understand that there is a program running on the Pico:

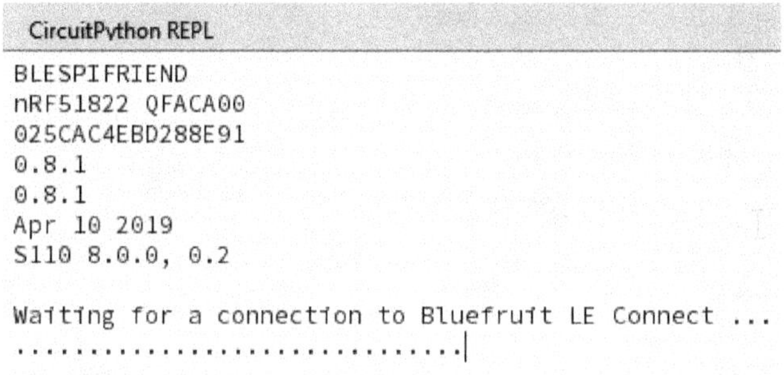

```
CircuitPython REPL

BLESPIFRIEND
nRF51822 QFACA00
025CAC4EBD288E91
0.8.1
0.8.1
Apr 10 2019
S110 8.0.0, 0.2

Waiting for a connection to Bluefruit LE Connect ...
.............................|
```

Figure 8.5 – Waiting for a connection

7. When a device is connected to our Bluetooth module, we print a debug message to the serial terminal (as shown in *Figure 8.6*):

```
    print("\n *Connected!*")
    connection_timestamp = time.monotonic()
    while True:
        if time.monotonic() - connection_timestamp > 3:
            connection_timestamp = time.monotonic()
            if not bluefruit.connected:
                break
        resp = bluefruit.uart_rx()
        if not resp:
            continue
```

Now, we wait for any incoming messages from the connected device. We also ensure that we are still connected to the device in question. If not, we break out of the loop and await connection from another device:

```
code.py output:
Initializing the Bluefruit LE SPI Friend module
BLESPIFRIEND
nRF51822 QFACA00
025CAC4EBD288E91
0.8.1
0.8.1
Apr 10 2019
S110 8.0.0, 0.2

Waiting for a connection to Bluefruit LE Connect ...
.............
 *Connected!*
```

Figure 8.6 – The Bluetooth module connected to the device

8. In the previous code snippet, we read messages using the uart_rx() method. If there is an incoming message to the Bluetooth module, we print it to the serial terminal. We also echo back the message in reverse using uart_tx():

```
print ("Read %d bytes: %s" % (len(resp), resp))
print ("Writing reverse...")
send = []
for i in range(len(resp), 0, -1):
    send.append(resp[i-1])
print(bytes(send))
bluefruit.uart_tx(bytes(send))
```

9. When we save the code to the Pico as code.py, interface the Bluetooth module, and execute the code, we should be able to see the following output, where the Bluetooth module is initialized and waiting for a connection:

```
CircuitPython REPL

BLESPIFRIEND
nRF51822 QFACA00
025CAC4EBD288E91
0.8.1
0.8.1
Apr 10 2019
S110 8.0.0, 0.2

Waiting for a connection to Bluefruit LE Connect ...
.................................|
```

Figure 8.7 – The Bluetooth module initialized and waiting for a connection

10. Download **Bluefruit LE Connect** from Adafruit on your smartphone. It is available for both Android and Apple devices. A screenshot of the application is shown in the following figure:

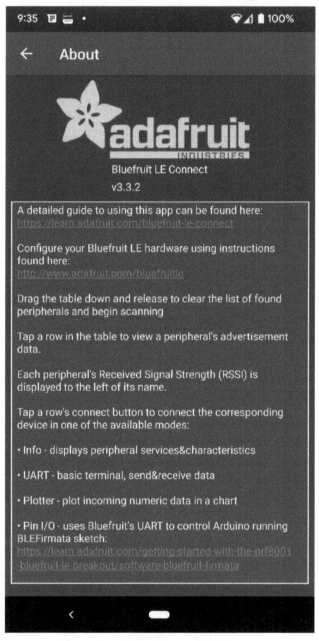

Figure 8.8 – Adafruit Bluefruit LE Connect

11. When you launch the application on your phone, you should be able to see a screen that lists the devices in your vicinity. You should also be able to spot **PicoBLE**, as shown in the following screenshot:

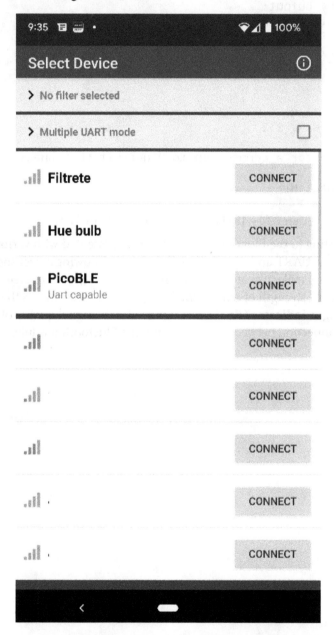

Figure 8.9 – PicoBLE detected on an Android device

12. When we connect to the device, we should be able to see the following output on the serial terminal of the Pico:

```
code.py output:
Initializing the Bluefruit LE SPI Friend module
BLESPIFRIEND
nRF51822 QFACA00
025CAC4EBD288E91
0.8.1
0.8.1
Apr 10 2019
S110 8.0.0, 0.2

Waiting for a connection to Bluefruit LE Connect ...
. . . . . . . . . . . . .
  *Connected!*
```

Figure 8.10 – The Bluetooth module interface of a Pico

13. After connecting to the Bluetooth module, we are presented with various module options such as UART and Plotter (as shown in the following screenshot). For this example, we will make use of the UART interface. The following screenshot also shows the signal strength of our *PicoBLE* module. This is known as the **Received Signal Strength Indicator** (**RSSI**), and it indicates the signal quality of the Bluetooth connection between the phone and the Bluetooth module:

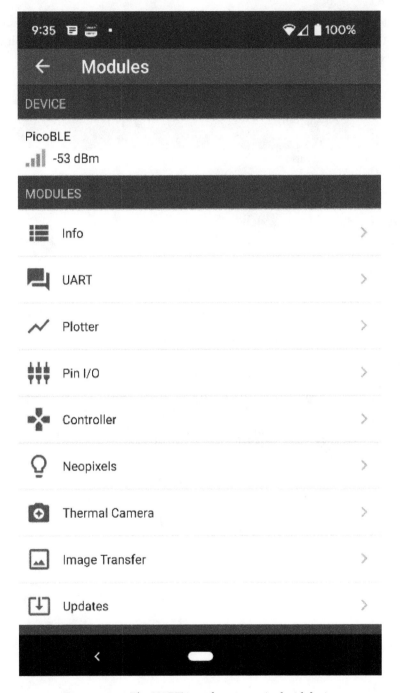

Figure 8.11 – The UART interface on an Android device

14. After selecting the UART interface, we will transmit the **Hello World** message to the Bluetooth module (as shown in the following screenshot). Note that the Bluetooth module echoes back the message in reverse:

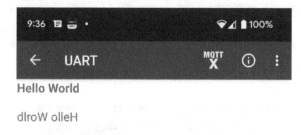

Figure 8.12 – Transmitting a message to the Bluetooth module and receiving a response

15. On the Pico, we will see debug messages on the serial terminal. In the following screenshot, note that the Bluetooth module has received 12 bytes (`Hello World\n`) and echoes back the message in reverse:

```
Waiting for a connection to Bluefruit LE Connect ...
. . . . . . . . . . . . . .
 *Connected!*
Read 12 bytes: b'Hello World\n'
Writing reverse...
b'\ndlroW olleH'
```

Figure 8.13 – Debug messages on the Pico

Now, we have connected our Pico to a mobile device using a Bluetooth module. In the next section, we will discuss publishing sensor data to the phone that could be published to the cloud using Adafruit IO.

Publishing sensor data via the Bluetooth module

In this section, we will discuss publishing data collected using the Bluetooth module to Adafruit IO. If you recall, we used this service to publish data to the cloud in *Chapter 3, Home Automation Projects*, and hence, we recommend following the instructions in that chapter to sign up for an account for the service. The following diagram shows the flow of data from your Pico to Adafruit IO:

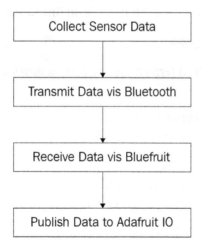

Figure 8.14 – Publishing data to Adafruit IO

The data collected by the Pico is transmitted via Bluetooth if it is connected. The data collected by the smartphone is then published to Adafruit IO.

Considering an example scenario

Let's consider a scenario where we have installed carbon dioxide sensors or temperature sensors (interfaced with the Raspberry Pi Pico) across different locations for data collection. Now, we want to collect data from the sensors using a mobile device whenever we are in the vicinity of the sensor and publish data to the cloud.

In this section, we are going to discuss publishing the data collected from the sensor using the mobile application. We recommend following instructions from *Chapter 7, Designing a Visual Aid for Tracking Air Quality*, on interfacing the carbon dioxide sensor. You are welcome to choose any sensor. Alternatively, you can just send randomly generated data for testing purposes.

Programming the Pico

We are going to discuss the code sample needed to collect data from the carbon dioxide sensor and transmit it via Bluetooth when connected to a mobile device. The code sample discussed in this section is identical to the one discussed in the previous section. So, we will only discuss the differences between the two examples. The code sample discussed in this section is available for download in this chapter's GitHub repository as `bluefruit_SCD30.py`.

The steps involved in transmitting the carbon dioxide sensor data are as follows:

1. Apart from importing the requisite modules discussed in the previous section, we also import the `adafruit_scd30` module on line number 6:

    ```
    import adafruit_scd30
    ```

2. Now, we initialize the I2C interface and the carbon dioxide sensor on lines 14 and 15 respectively:

    ```
    i2c = busio.I2C(board.GP9, board.GP8)
    scd = adafruit_scd30.SCD30(i2c)
    ```

3. The other major change is between lines 49 and 52. If we are connected to a Bluetooth device and new data is available from a carbon dioxide sensor, we transmit the data using the `uart_tx` method:

    ```
    if scd.data_available:
        data = str(scd.CO2) + "\n"
            bluefruit.uart_tx(bytes(data, "utf-8"))
    time.sleep(2.0)
    ```

4. Now, we should be able to see the data on the phone, as shown in the following screenshot:

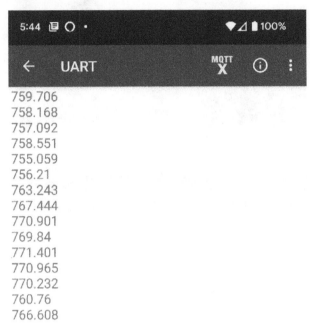

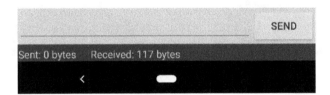

Figure 8.15 – The CO2 sensor data on a mobile device

5. Now, we should also be able to plot the collected data from the **Plotter** tab, as shown in the following screenshot:

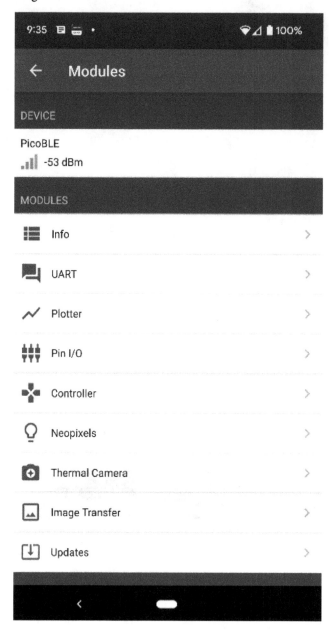

Figure 8.16 – The Plotter tab in the Bluefruit application

6. We should be able to see a live plot of the incoming data, as shown in the following screenshot:

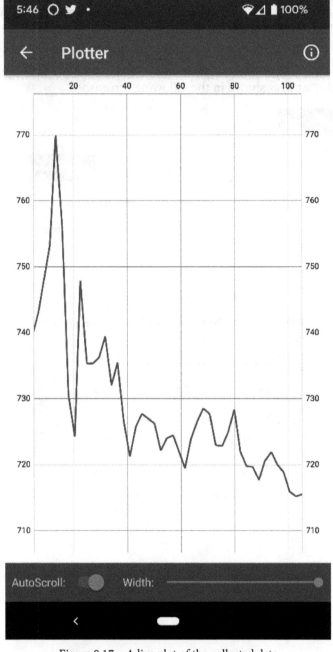

Figure 8.17 – A live plot of the collected data

7. Now, let's take a look at publishing the collected data using MQTT. **MQTT** stands for **MQ Telemetry Transport**. "MQ" refers to a product series developed by IBM. MQTT is a protocol developed for low power devices like the Raspberry Pi Pico to publish data to the cloud. We assume that you have signed up for an account at io.adafruit.com following the instructions from *Chapter 3, Home Automation Projects*.

8. Go to io.adafruit.com and create a new feed (from the **Feeds** tab) called Collected Data, as shown in the following screenshot:

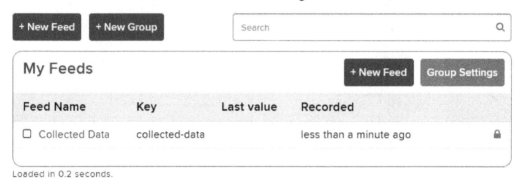

Loaded in 0.2 seconds.

Figure 8.18 – Creating a new data feed

9. You can find the feed information under **Feed Info**, as shown in the following screenshot:

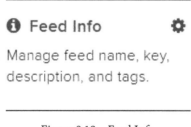

Figure 8.19 – Feed Info

10. We need the MQTT feed name, as shown in the following screenshot:

Current Endpoints

Web	https://io.adafruit.com/	'feeds/collected-data
API	https://io.adafruit.com/api/v2	eeds/collected-data
MQTT by Key	/feeds/collected-data	

Figure 8.20 – The MQTT feed name

11. Now, we get back to our mobile application and tap the **MQTT** button (as shown in the following screenshot):

Figure 8.21 – Tap the MQTT button

12. On this page, we will enter the MQTT feed name in the **Uart RX** field, as shown
in the following screenshot. We also need to provide the key for our Adafruit IO
account. This information is available under the **My Key** tab of the account. There
is a *QR code* on the page that enables you to capture the credentials using your
smartphone's camera. When you tap the **Scan Code** button, it launches the camera
application of your smartphone. When you scan the QR code, the **Pass/Key** field is
filled automatically. You just need to enter your username. We're not showing the
My Key tab of the Adafruit IO page for privacy reasons:

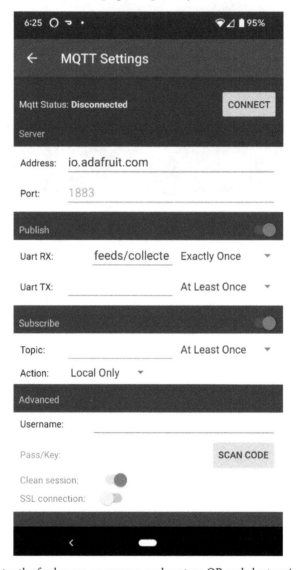

Figure 8.22 – Enter the feed name, username, and capture QR code by tapping SCAN CODE

13. Now, tap the **Connect** button (as shown in the previous screenshot), and you should connect to Adafruit's MQTT server. The incoming data is automatically published to the MQTT server.

14. You should be able to view the data from the **Feeds** page of your Adafruit IO account:

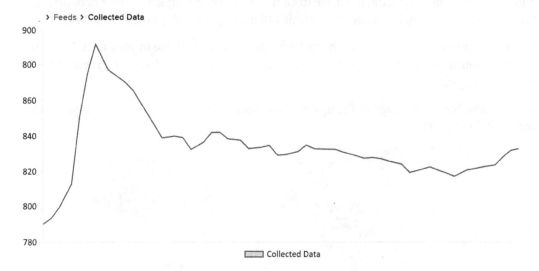

> Feeds > Collected Data

Figure 8.23 – Collected carbon dioxide sensor data

Now, you can build nodes and collect data using the Bluetooth module. In the next section, we will review interfacing the Raspberry Pi Pico with a Sigfox module.

Interfacing a Sigfox module

In this section, we will discuss interfacing a Sigfox module with the Raspberry Pi Pico. Before we get started, let's take a quick look at Sigfox.

What is Sigfox?

Sigfox is a type of **Low-Power Wide Area Network** (**LPWAN**) that is operated by a French company of the same name. Sigfox radios have a range of about 10–30 km, and the network operator restricts the total number of transmissions to 140 messages per day with each message restricted to 12 bytes. This type of network is suitable for battery-powered wireless nodes deployed at scale. Sigfox typically operates in unlicensed ISM radio bands **ISM** refers to **Industrial, Scientific and Medical** bands. As the acronym suggests, it is meant to be used in those applications without a license. In the US, it is typically 915 MHz. You can learn more about Sigfox here: `https://build.sigfox.com/sigfox`.

> **Sigfox's future**
>
> At the time of writing this chapter, Sigfox filed for bankruptcy protection. We
> don't know what this means for the future of Sigfox networks.

Sigfox networks need an annual license from the operator to connect to their network. In
the United States, a single license to transmit 140 messages per day costs about 19 USD.

It is also important to find out whether a Sigfox network is available in your area. You can
find out whether coverage is available in your area here: `https://www.sigfox.com/`
`en/coverage`.

For example, Sigfox coverage in the upstate New York area of the United States is shown in
the following figure:

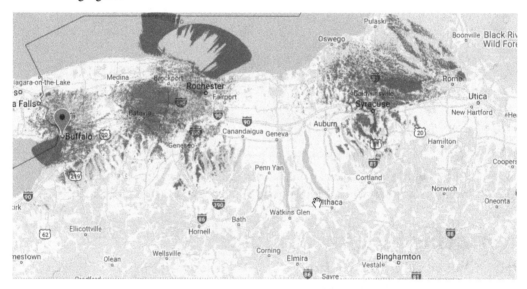

Figure 8.24 – Sigfox coverage in upstate New York

In the next section, we will review the Sigfox module used in this example.

Sigfox module

In this section, we will discuss the RC1692 Sigfox module used in this example (shown in
the following photo). RC1692 (available from `https://bit.ly/3DtRIap`) is a Sigfox
radio that operates in the 902–928 MHz frequency band. It is an FCC-certified module
that is meant to be used in the United States and a few other countries (you can find more
information from `https://bit.ly/32Wr9xX`).

We couldn't find any cheap development kits for Sigfox radio modules, so we designed our own **Printed Circuit Board (PCB)** in the Adafruit **Featherwing** form factor. You can build your own using the design files available here: `https://bit.ly/3pxf3mq`.

> **Pico-Compatible Add-On Board**
>
> While this example uses the Featherwing, a board that is compatible with the Pico is in the works, which could be used for future project. We will upload the design files to this book's repository.

We designed the board in the Featherwing factor because we built the radio before the arrival of the Raspberry Pi Pico, but interfacing the module is quite simple since it comes with a UART interface.

Our Sigfox Featherwing also comes with a six-pin header meant for a standard FTDI adapter (such as the one at `https://bit.ly/32Ws15H`). This enables you to register the radio with the Sigfox network. The following photo shows the Sigfox Featherwing board sitting on top of a breadboard:

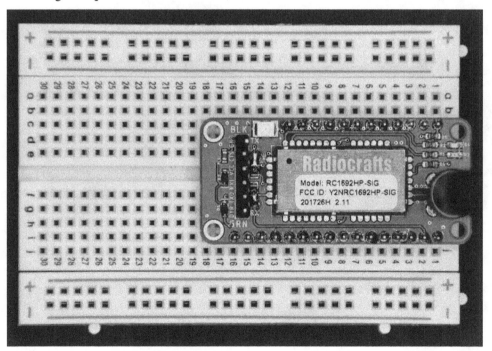

Figure 8.25 – RC1692 Sigfox Featherwing

In the next section, we will discuss setting up the Sigfox radio before we interface it with the Raspberry Pi Pico.

Setting up the Sigfox module

In this section, we will discuss setting up the Sigfox module. The steps include the following:

1. The first step is to ensure that there is a Sigfox network operator in your area. You can find the coverage information in your area using the link shared earlier in this section.

2. If Sigfox is available in your area, you can buy an annual license from `https://buy.sigfox.com/`. The pricing depends upon the operator and options available in your area. In the United States, a license to transmit 140 messages per day costs 19 USD per year. There is also an alternative plan available for 8 USD per year, which is restricted to two messages per day.

3. You also need to create an account at `https://backend.sigfox.com/auth/login`. The purchased license is automatically added to your Sigfox account.

4. Now, you can create an account from the tab at the top of the web page, as shown in the following screenshot:

Figure 8.26 – Adding a new device to your Sigfox backend

5. When you click on **New**, it will take you to a page where you can add information related to the new device.

6. We will register the device as a prototype, and we also need some unique identifiers belonging to the module. We will make use of the *SIG-CCT* software from the manufacturer's website: `http://www.radiocrafts.com/`.

7. We will follow the guide available from the manufacturer to retrieve the information needed to register the module (`https://bit.ly/3rDMXsw`).

8. Once we have registered the device, it is time to wire up the Sigfox Featherwing to the Pico. The connections to the module are as follows, where the left-hand side refers to a pin on the Raspberry Pi Pico while the right refers to one on the Sigfox *Featherwing* board:

   - 5V → VCC
   - GP4 (TX) → RX
   - GP5 (RX) → TX
   - GP19 → RESET
   - GND pins tied together

The following photo shows the pins used to interface the Sigfox Featherwing with the Raspberry Pi Pico:

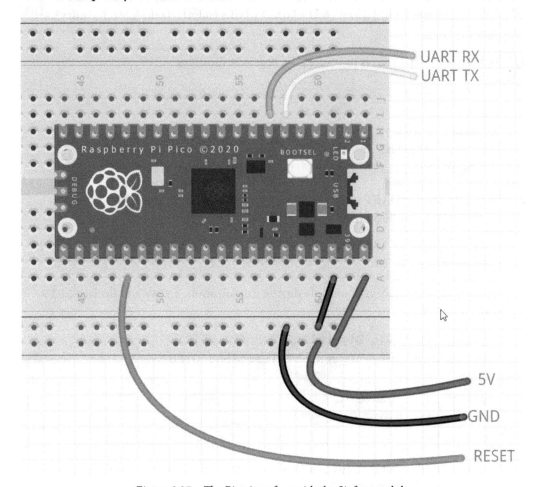

Figure 8.27 – The Pico interface with the Sigfox module

In the next section, we are going to discuss the code needed to control and transmit data using the Sigfox module.

Discussing the code sample

Let's review the code needed to control the Sigfox module. The code discussed here is available for download in the `downloads` folder of this chapter's GitHub repository as `uart_sigfox.py`:

1. As always, the first step is to import the requisite modules:

    ```
    import time
    import board
    import busio
    import digitalio
    import binascii
    ```

2. Since we need the UART interface, we are going to initialize the UART interface using the `GP4` and `GP5` GPIO pins, as follows:

    ```
    uart = busio.UART(tx=board.GP4, rx=board.GP5,
    baudrate=19200)
    ```

3. We need to reset the Sigfox module during initialization. So, we will declare `GP19` as an output pin:

    ```
    pin = digitalio.DigitalInOut(board.GP19)
    pin.direction = digitalio.Direction.OUTPUT

    pin.value = False
    time.sleep(5.0)
    pin.value = True
    ```

4. Now, we will transmit data twice with a 15-second delay between consequent transmissions. Each transmission needs to begin with the length of the payload. In the following code snippet, the first element of `data` refers to the length – for example, 2 bytes. We are sending 2 random bytes of `0x10` and `0x20`:

    ```
    data = [2, 0x10, 0x20]
    uart.write(bytes(data))
    time.sleep(10)
    uart.write(bytes(data))
    print("Done")
    ```

Sigfox Message Restrictions

As mentioned earlier, Sigfox restricts to 140 messages per day. Do not use an infinite loop and let your code run forever. A Sigfox wireless node is supposed to send messages once or twice a day.

5. When we put the preceding code sample together, we get the following:

```
import time
import board
import busio
import digitalio
import binascii


pin = digitalio.DigitalInOut(board.GP19)
pin.direction = digitalio.Direction.OUTPUT

uart = busio.UART(tx=board.GP4, rx=board.GP5,
baudrate=19200)
data = [2, 0x10, 0x20]


pin.value = False
time.sleep(5.0)
pin.value = True


uart.write(bytes(data))
time.sleep(10)
uart.write(bytes(data))
print("Done")
```

When we save the preceding code sample as `code.py` and run it, we will be able to transmit data to the Sigfox network. Whenever the data is transmitted successfully, the blue LED on the Sigfox module blinks a couple of times. We can view the data on the Sigfox backend, as shown in the following screenshot, where the two random bytes transmitted can be viewed on the dashboard:

	page 1				
Time	Seq Num	Data / Decoding	LQI	Callbacks	Location
2021-12-06 13:19:02	185	1020			
2021-12-06 13:18:46	184	1020			

Figure 8.28 – Transmitted data on the Sigfox dashboard

You can view the coarse location of the data transmission by clicking on the location pin. In this case, the data is being transmitted from Buffalo, New York. Now, we recommend interfacing a sensor of your choice and publishing the data. Keep in mind that you only have 12 bytes per message:

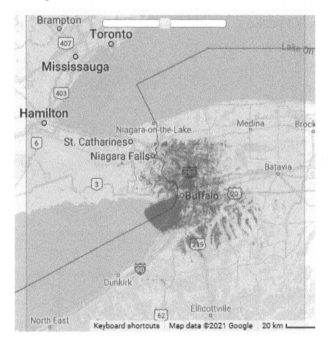

Figure 8.29 – The location of the data transmission

In the next section, we will discuss building LoRa nodes using Raspberry Pi Pico.

Interfacing a LoRa module

In this section, we are going to interface two LoRa modules with the Raspberry Pi Pico. This will enable us to establish communications between two wireless nodes.

What is LoRa?

Before we get started with our example, let's get started with a brief introduction to LoRa. **LoRa** stands for **Long Range Radio** and is a wireless protocol meant for long-range and low-power communications. It was developed by an organization called *Semtech*. While LoRa and Sigfox (discussed in the previous section) are similar, the main difference is that the former lets you set up your own network, while the latter is operated by an individual operator that requires a license to connect to their network. The following table highlights the difference between LoRa and Sigfox radios:

LoRa	Sigfox
You can set up your own network in any location of your choice.	A licensed operator operates the network. Restricted to where Sigfox networks operate.
You need to set up your own gateway and backend infrastructure. No need to pay an annual license fee.	Sigfox operates an annual license. The message can be routed from Sigfox's backend to your cloud solution.
No restrictions on the number of messages per day.	Restricted to 140 messages per day.

Table 8.1 – The differences between LoRa and Sigfox

There is a nonprofit organization called the LoRa Alliance that has developed a standard for interoperability between various LoRa networks. LoRa radios make use of the license-free ISM bands, which are on 915 MHz in the United States. You can learn more about LoRa here: `https://bit.ly/3rNy1YS`.

Considering an example scenario

Let's consider a scenario where you are going to install a sensor in a location where you don't necessarily have a Wi-Fi network. This location doesn't have any buildings in the vicinity, so you need a battery-powered device. This is where LoRa modules can be suitable. You can also build wireless nodes enabled by a LoRa module. While one node is installed at the location, the other one can be installed indoors. The second node can also be interfaced to a Wi-Fi module (as shown in the following diagram). This enables you to collect sensor data and publish it to the cloud using the Wi-Fi module:

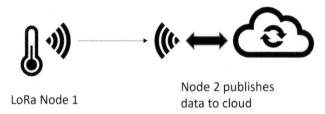

LoRa Node 1   Node 2 publishes
                               data to cloud

Figure 8.30 – The flow of data between two LoRa nodes

Let's get started.

In this example, we will make use of two Raspberry Pi Pico boards and two RFM95W LoRa Radio Transceiver Breakout boards. Our first step is to interface each RFM95W breakout with the Raspberry Pi Pico.

The RFM95W is a LoRa radio capable of transmitting up to a 2 km line of sight. You can learn more about the radio here: `https://bit.ly/3BiR7Ja`.

Before we get started, some assembly of the RFM95W Breakout board is required. We recommend following the instructions here to prepare it: `https://bit.ly/339Zwlh`.

The breakout board is connected to the Pico as follows, where the left side of the arrow refers to a pin on the Pico while the right side refers to a pin on the RFM95W breakout:

- GP5 → CS
- GP6 → RESET
- GP2 → SCK
- GP3 → MOSI
- GP4 → MISO
- VBUS → VIN
- GND

The following figure shows the Fritzing schematic to interface the RFM95W Breakout with the Raspberry Pi Pico:

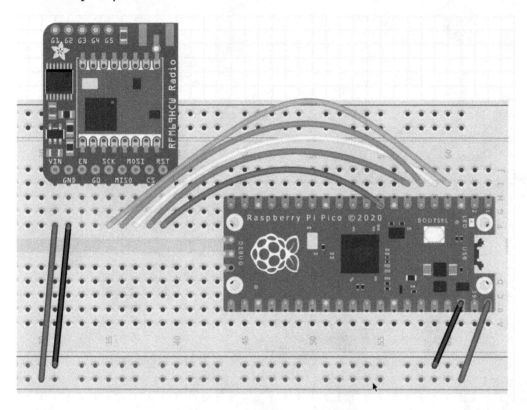

Figure 8.31 – The schematic to interface the RFM95W Breakout with the Raspberry Pi Pico

A breadboard setup of the wireless node is shown in the following photo:

Figure 8.32 – The breadboard setup of the wireless node

Let's look at the code needed to transmit the message using the RFM95W Breakout. The code discussed in this section is available for download in the downloads folder of this chapter's GitHub repository as code_lora.py:

1. The first step in the code is to import the requisite modules:

```
import board
import busio
import digitalio
import adafruit_rfm9x
```

2. The next step is to declare the radio frequency. Since we are using a 915 MHz radio breakout, we will define the frequency using the `RADIO_FREQ_MHZ` variable. You need to change this according to the approved frequency for LoRa modules in your country – for example, the frequencies for LoRa modules are either 433 or 863 MHz:

```
RADIO_FREQ_MHZ = 915.0
```

3. Now, we will initialize the SPI interface and the radio breakout:

```
CS = digitalio.DigitalInOut(board.GP5)
RESET = digitalio.DigitalInOut(board.GP6)

spi = busio.SPI(board.GP2, MOSI=board.GP3, MISO=board.GP4)
rfm9x = adafruit_rfm9x.RFM9x(spi, CS, RESET, RADIO_FREQ_MHZ)
```

4. We will use the onboard LED to indicate when there is an incoming packet:

```
LED = digitalio.DigitalInOut(board.GP25)
LED.direction = digitalio.Direction.OUTPUT
```

5. The next step is to set the transmission power:

```
rfm9x.tx_power = 23
```

6. Now, we enter the main loop and make a test transmission. After the test transmission, we wait for any incoming messages. If there are no incoming messages, we turn off the LED. If there are incoming messages, we decode them and print them on the screen:

```
while True:
    rfm9x.send(bytes("Hello world!\r\n", "utf-8"))
    packet = rfm9x.receive()

    if packet is None:
        LED.value = False
        print("Received nothing! Listening again...")
    else:
        LED.value = True
        print("Received (raw bytes): {0}".format(packet))
```

```
            packet_text = str(packet, "ascii")
            print("Received (ASCII): {0}".format(packet_
text))


            rssi = rfm9x.last_rssi
            print("Received signal strength: {0} dB".
format(rssi))
```

7. When we save the preceding code sample as code.py on both the Picos, we get the
    following output whenever we receive a message from each Pico:

```
Received (raw bytes): bytearray(b'Hello world!\r\n')
Received (ASCII): Hello world!

Received signal strength: -49 dB
Received nothing! Listening again...
Received nothing! Listening again...
Received (raw bytes): bytearray(b'Hello world!\r\n')
```

Figure 8.33 – The message received via the LoRa breakout

If you interface a sensor with the Pico, as discussed in the Bluetooth example, you should
be able to collect data using one wireless node and transmit it to the other. We will share
an example in this book's repository. If the other node is interfaced with a Wi-Fi module,
you can publish the collected data to a service such as *ThingSpeak* or *Adafruit IO*.

Now, if you have multiple nodes collecting data, you can install a LoRa gateway to collect
data from all the wireless nodes. We have provided more information in this book's
GitHub repository.

Summary

In this chapter, we got started by interfacing a Bluetooth module with a Raspberry Pi Pico.
Then, we discussed publishing sensor data via Bluetooth using a mobile device. We also
interfaced a Sigfox module with the Pico. Finally, we built two LoRa nodes using the Pico.

In the next chapter, we will discuss building a robot using the Pico.

9

Let's Build a Robot!

In this chapter, we will have some fun building a robot using a Raspberry Pi Pico. We will discuss programming the Raspberry Pi Pico using MicroPython. We will also discuss building a robot using an off-the-shelf robotics kit. We will get started by reviewing the components of our robot by testing them; then, we will discuss building a line-following robot and an obstacle avoiding robot.

Figure 9.1 – Kitronik Autonomous Robotics Platform for Raspberry Pi Pico

This can be a fun weekend project that could be further enhanced to participate in robotic contests. We will give you some ideas toward the end of this chapter.

Figure 9.2 – Line-following robot using the Kitronik kit

The topics discussed in this chapter include the following:

- Installing the prerequisites
- Controlling the LEDs
- Motor selection and control
- Testing the sensors
- Testing the robot

Technical requirements

The following hardware is recommended for this chapter:

- Raspberry Pi Pico (`https://www.adafruit.com/product/4883`) – USD 4
- Kitronik Autonomous Robotics Platform (`https://kitronik.co.uk/products/5335-autonomous-robotics-platform-for-pico`) – USD 45

The code samples for this chapter are available at `https://github.com/PacktPublishing/Raspberry-Pi-Pico-DIY-Workshop/tree/main/chapter_09`.

Code in Action videos for this chapter can be viewed at `https://bit.ly/3kNCFkM`.

In the next section, we will install the required library for our robot.

Installing the prerequisites

Before we get started, we need to install the Thonny IDE, along with installing MicroPython on the Raspberry Pi Pico.

> **MicroPython Installation**
>
> We recommend installing MicroPython on your Pico. If you are not familiar with the installation process, you can follow the installation process in *Chapter 1, Getting Started with the Raspberry Pi Pico.*

The required library for the Kitronik Robotics kit is available for download as a ZIP file from its repository at `https://bit.ly/3K35aoM`.

Once the contents of the files are extracted, we can install the library via the Thonny IDE as follows:

1. In the Thonny IDE, go to the folder containing the extracted library files, as shown in the following screenshot, where you can navigate to the location of the files:

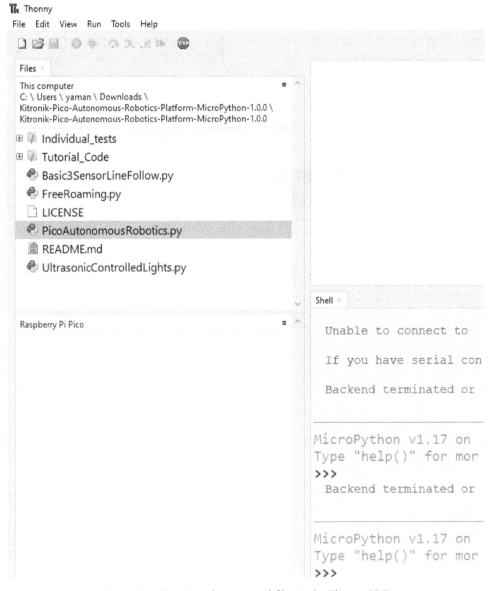

Figure 9.3 – Locating the extracted files in the Thonny IDE

2. Now, right-click on `PicoAutonomousRobotics.py`, and upload it to the Pico, as shown in the following screenshot:

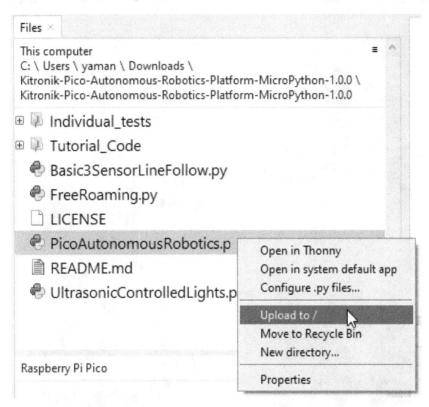

Figure 9.4 – Uploading the library to the Pico

3. Now that the library is installed on the Pico, we will take it for a spin in the upcoming sections. In the following screenshot, you will notice the library copied over to the Pico:

Figure 9.5 – Library copied over to the Pico

In the upcoming sections, we will test the various features of the library. First, we need to install the batteries for our project.

Installing the batteries

We are assuming you have installed the batteries, as shown in the following figure. The robot needs four AA batteries.

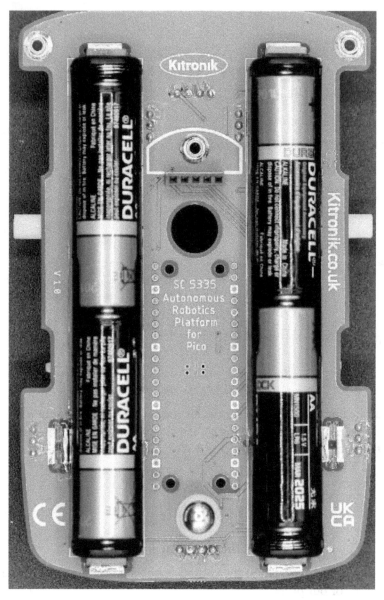

Figure 9.6 – Batteries installed

Ensure that the sliding switch is set to the *OFF* position, as shown in the following figure:

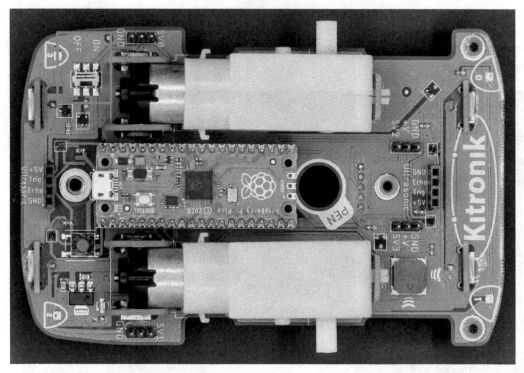

Figure 9.7 – Switching the location on the Robotics Platform

We will get started by testing the RGB LEDs in the next section.

Controlling the LEDs

In this section, we will get started with testing the RGB LEDs on the Kitronik Robotics kit. There are four RGB LEDs on the chassis (highlighted in *Figure 9.6*). The LEDs could be used to provide a visual indication of the robot's action. We will light them up in a circular pattern.

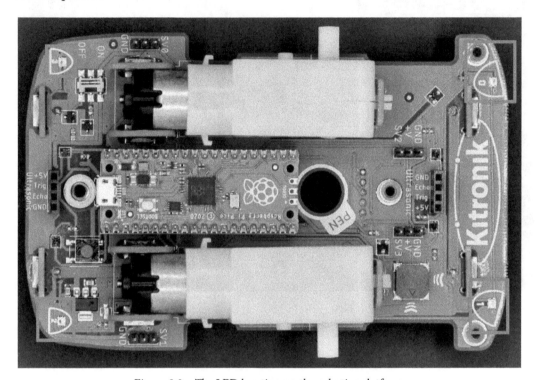

Figure 9.8 – The LED location on the robotics platform

The code sample discussed in this section is `rgb_led_test.py`:

1. In the Thonny IDE, create a file called `main.py` and save it to your Pico, as shown in the following figure, where the dialog appears when you click on the **Save** button.

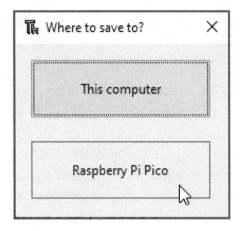

Figure 9.9 – Save main.py to Raspberry Pi Pico

2. Now, let's discuss the code meant to drive the LEDs in a circular pattern. The first step is to import the requisite modules. We are going to need the KitronikPicoRobotBuggy class from the library we installed earlier, along with the time module:

```
from PicoAutonomousRobotics import KitronikPicoRobotBuggy
from time import sleep
```

3. The next step is to initialize an object belonging to the KitronikPicoRobotBuggy class. We will also initialize the state variable called LEDState. This is used to rotate the LED colors between the different states:

```
my_robot = KitronikPicoRobotBuggy()
LEDState = 0
```

4. Now, we declare a function called LEDPattern(), where we switch the LED color based on the LEDState state variable. We use the setLED method to set the LED color. This method accepts two arguments, namely, the LED position and the color. Refer to *Figure 9.6* for the LED position. The rectangle highlights the LED and the position number:

```
def LEDPattern():
    global LEDState
    if LEDState == 0:
        my_robot.setLED(0,my_robot.BLUE)
        my_robot.setLED(1,my_robot.GREEN)
        my_robot.setLED(2,my_robot.YELLOW)
```

```
        my_robot.setLED(3,my_robot.RED)
        LEDState = 1
```

5. In the previous code snippet, we set the LED colors and set the `LEDState` variable to 1. When `LEDState` is 1 during the subsequent execution, we set the LED colors as shown in the following snippet, and change the `LEDState` state variable to 2:

```
    elif LEDState == 1:
        my_robot.setLED(0,my_robot.BLUE)
        my_robot.setLED(1,my_robot.GREEN)
        my_robot.setLED(2,my_robot.YELLOW)
        my_robot.setLED(3,my_robot.RED)
        LEDState = 2
```

6. Before exiting the function, we will call the `show()` method to display the new LED colors:

```
    my_robot.show()
```

7. In our `while` loop, we will call the `LEDPattern()` function with a 1 second delay:

```
    while True:
        LEDPattern()
        sleep(1)
```

8. When you click on the **Run** button, you should notice the LEDs changing colors in a circular pattern. The following figure shows the LED colors. We must note that the LED colors don't pop because of the lighting conditions.

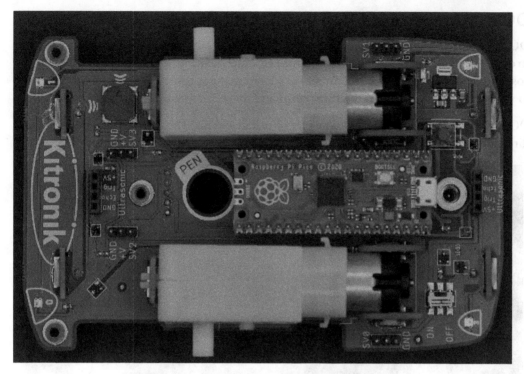

Figure 9.10 – LEDs changing colors on the platform

When we save the code sample and execute it (by pressing the **Run** button), you will notice the LEDs changing colors in a circular pattern (as shown in the previous figure). Now that we have tested the RGB LEDs, we will discuss the motor.

Motor selection and control

In this section, we will discuss the various motor options available to build a robot. For a typical desktop robot, the various motor options include the following:

- Direct Current (DC) motors
- Stepper motors
- Servo motors

Let's discuss the applications of these motors.

DC motors

DC motors are ideal for wheeled robots. The following figure shows the DC motors used to drive the wheels of our robot. DC motors are widely used in toys because of their low cost and their small size. The DC motors used in our robot come with an in-built gearbox to reduce the **revolutions per minute (rpm)** and make it suitable for a wheeled robot.

DC motors are typically operated using an H-bridge circuit. This enables the speed and direction of travel to be controlled.

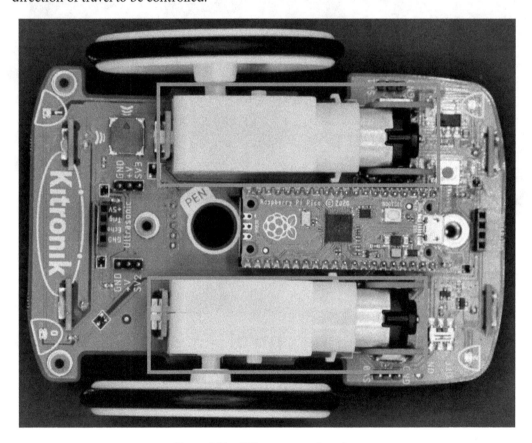

Figure 9.11 – DC motors on the robot

Stepper motors

Stepper motors are ideal for positioning applications. For example, stepper motors can be used to drive linear mechanisms, such as a threaded shaft, and position the end effector at the required position. Stepper motors are ideal for applications such as 3D printing, and they consist of pairs of coils that are driven by energizing the individual coils. In the following figure, two stepper motors are used to drive a maze-solving toy:

Figure 9.12 – Maze-solving robot

Next, we will discuss servo motors.

Servo motors

Servo motors are typically controlled by a pulse. The length of the pulse determines the angle of rotation. Servo motors are typically used in applications such as a gripper (shown in the following figure), where a pulse signal is used to hold or release an object:

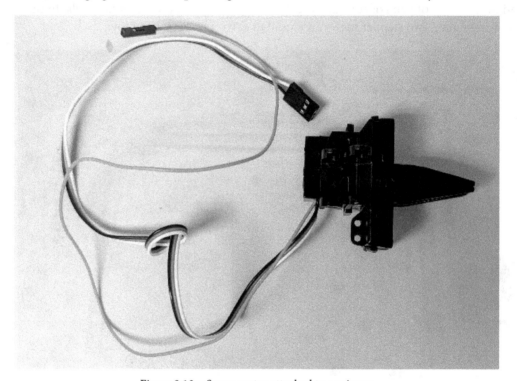

Figure 9.13 – Servo motor attached to a gripper

In the preceding figure, the servo motor comes with a fourth cable that provides voltage feedback on the position. This helps with correcting the servo motor position. The robotics platform that we chose for this project can drive up to four servo motors. The servo motor pins are highlighted in the red rectangles shown in the following figure:

Figure 9.14 – Servo motor pins

For a further look at the various motors and their working principles, we recommend the following article: `https://learn.sparkfun.com/tutorials/motors-and-selecting-the-right-one/all`.

Now that we have discussed the various motors available to build a robot, let's discuss motor control to drive the robot.

DC motor control

In this section, we are going to test the DC motor used to drive the robot by moving it forward and backward for 10 seconds. The code sample discussed in this section is available for download along with this chapter as `dc_motor_test.py`.

Before we get started, ensure that the power supply to the robot is set to *OFF* (as shown in *Figure 9.6*). This is to ensure that we don't have a runaway robot. Alternatively, you can remove the wheels of the robot for testing purposes.

Let's get started:

1. In the Thonny IDE, create a file called `main.py` and save it to your Pico.

2. The first step is to import the requisite modules. We are going to need the `KitronikPicoRobotBuggy` class:

```
from PicoAutonomousRobotics import KitronikPicoRobotBuggy
from time import sleep
```

3. The next step is to initialize an object belonging to the `KitronikPicoRobotBuggy` class. We will also initialize the state variable called `state`:

```
buggy = KitronikPicoRobotBuggy()
state = 0
```

4. When the `state` variable is `0`, we move the motor in the forward direction. We use the `motorOn` method to rotate the motor. The method accepts three arguments, namely, the motor (left `"l"`, or right `"r"`), direction (forward `"f"`, or reverse `"r"`), and speed (a value between 0 and 100):

```
def MotorControl():
    global state

    if state == 0:
        buggy.motorOn("l","f",100)
        buggy.motorOn("r","f",100)
        state = 1
```

5. We can turn the motor off using the `motorOff` method. We just need to specify the motor that needs to be turned off (left `"l"` or right `"r"`):

```
buggy.motorOff("l")
```

```
buggy.motorOff("r")
state = 0
```

6. Putting it all together, we have the following:

```
from PicoAutonomousRobotics import KitronikPicoRobotBuggy
from time import sleep

buggy = KitronikPicoRobotBuggy()
state = 0

def MotorControl():
    global state

    if state == 0:
        buggy.motorOn("l","f",100)
        buggy.motorOn("r","f",100)
        state = 1
    elif state == 1:
        buggy.motorOn("l","r",100)
        buggy.motorOn("r","r",100)
        state = 2
    else:
        buggy.motorOff("l")
        buggy.motorOff("r")
        state = 0

while True:
    MotorControl()
    sleep(10)
```

When we execute the code by clicking on the **Run** button, you will notice the motors moving forward and backward every 10 seconds.

In the next section, we will discuss driving the servo motors.

Servo motor control

In this section, we will discuss controlling the servo motors using the pins available on the robotics platform. As mentioned earlier, the platform comes with the capability to control up to four motors (as shown in *Figure 9.10*).

We are going to discuss controlling the gripper using the servo motor. The code sample discussed in this section is available for download along with this chapter as servo_ motor_test.py.

The servo motor that is connected to the gripper is connected to port 2 of the robotics platform, as shown in the following figure:

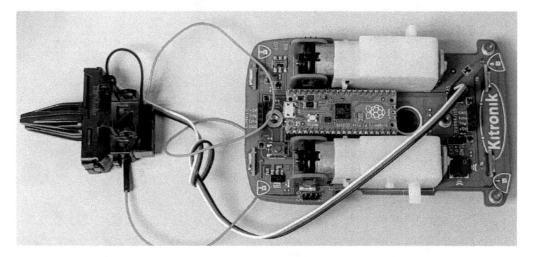

Figure 9.15 – Controlling the gripper using a servo motor

Let's discuss the code needed to open and close the gripper:

1. In the Thonny IDE, create a file called main.py and save it to your Pico.

2. The first step is to import the requisite modules. We are going to need the KitronikPicoRobotBuggy class:

    ```
    from PicoAutonomousRobotics import KitronikPicoRobotBuggy
    from time import sleep
    ```

3. The next step is to initialize an object belonging to the KitronikPicoRobotBuggy class:

    ```
    buggy = KitronikPicoRobotBuggy()
    ```

4. In order to fully close the gripper, we make use of the `goToPosition()` method to direct the servo motor to 180 degrees. The method takes two arguments, namely, the servo motor port number (the gripper servo motor is connected to port 2), and the angle of rotation:

```
buggy.goToPosition(2,180)
sleep(5)
```

The following figure shows the gripper in a fully closed position:

Figure 9.16 – Gripper in a fully closed position

5. In order to fully retract the gripper, we call the same method again by setting the servo position to 0 degrees:

```
buggy.goToPosition(2,0)
sleep(5)
```

The following figure shows the gripper in a fully retracted position:

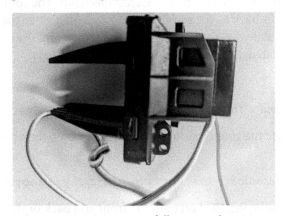

Figure 9.17 – Gripper in a fully retracted position

6.   Putting it all together, we have the following:

```
from PicoAutonomousRobotics import KitronikPicoRobotBuggy
from time import sleep

buggy = KitronikPicoRobotBuggy()

while True:
    buggy.goToPosition(2,180)
    sleep(5)
    buggy.goToPosition(2,0)
    sleep(5)
```

When we execute the code by clicking on the **Run** button, we can observe the gripper closing and retracting every 5 seconds.

In the next section, we will test the sensors that come with the robotics kit.

Testing the sensors

In this section, we will test the ultrasonic sensor and the line-following sensor that come with the robotics kit. We will start by testing the ultrasonic sensor.

Ultrasonic sensor

Ultrasonic sensors are generally used in obstacle-avoidance applications. They use the **time of flight** principle to measure the distance between objects. The sensor transmits a sound signal at a known frequency in the ultrasonic spectrum. The sound signal bounces off the surface of obstacles back to the sensor. The time elapsed since the transmission of the signal is used to calculate the distance between the sensor and the object. The distance between objects is calculated as follows:

Speed = Distance / Time

Distance = (Speed of Sound x Time) / 2

We divide by two because the sound has to travel twice between the obstacle and the sensor.

You can learn about ultrasonic sensors at `https://bit.ly/3xeqpS2`.

The ultrasonic sensor can be installed on both the front and back of the robot. For this example, we will install it on the front (as shown in the following figure):

Figure 9.18 – Ultrasonic sensor installed on the robot

The overhead view of the sensor installation is shown in the following figure:

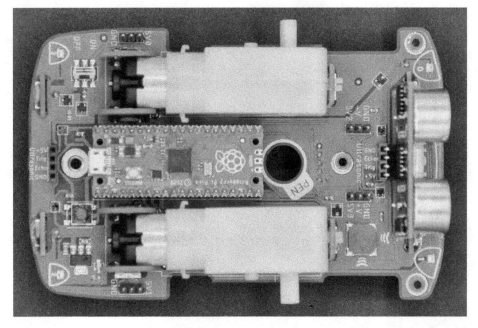

Figure 9.19 – Overhead view of the ultrasonic sensor

Let's review the code sample for testing the ultrasonic sensor. The code sample discussed in this section is available for download along with this chapter as `ultrasonic_test.py`:

1. In the Thonny IDE, create a file called `main.py` and save it to your Pico.

2. The first step is to import the requisite modules. We are going to need the `KitronikPicoRobotBuggy` class:

    ```
    from PicoAutonomousRobotics import KitronikPicoRobotBuggy
    from time import sleep
    ```

3. The next step is to initialize an object belonging to the `KitronikPicoRobotBuggy` class:

    ```
    buggy = KitronikPicoRobotBuggy()
    ```

4. Now, we can retrieve the distance in centimeters using the `getDistance()` method by specifying the sensor position. The method accepts `"f"` for the front sensor and `"r"` for the rear sensor:

    ```
    while True:
        frontDistance = buggy.getDistance("f")
        print(frontDistance)
        sleep(2)
    ```

5. We can use this distance information to control our robot. Putting it all together, we have the following:

    ```
    from PicoAutonomousRobotics import KitronikPicoRobotBuggy
    from time import sleep

    buggy = KitronikPicoRobotBuggy()

    while True:
        frontDistance = buggy.getDistance("f")
        print(frontDistance)
        sleep(2)
    ```

6. When we execute the code using the **Run** button and point the robot at various objects or use your hand to simulate an obstacle, we should see the following output where the printed values represent the distance to the obstacle:

```
MicroPython v1.18 on 2022-01-17; Raspberry Pi Pico with RP2040
Type "help()" for more information.
>>> %Run -c $EDITOR_CONTENT
  17.06425
  15.41785
  42.0861
  109.8457
  13.66855
  5.3165
  187.9983
```

Figure 9.20 – Ultrasonic sensor output

In the preceding figure, the distance to various objects is measured in centimeters. The sensor output can be noisy at times when a very high value such as 109 and 187 cm is displayed. This happens by placing fingers right next to the sensor. Therefore, it is important to filter out such values for reliable navigation of the robot.

Next, we will test the line-following sensor.

Line-following sensor

In this section, we are going to discuss testing the line-following sensors. As the name suggests, line-following sensors are used to make the robot follow a line. This line could be a dark line on a lighter background or vice versa. A line-following sensor typically consists of an infrared LED and a photodiode. The line-following sensor makes use of the reflectivity of surfaces to drive the robot. Lighter surfaces tend to reflect more light than darker surfaces.

Let's consider a scenario where we are driving the robot to follow a dark line on a lighter background. A typical line-following robot has three sensors, namely, left, right, and center. The idea is to drive the robot such that the center sensor stays on top of the dark line. When the robot is centered on the line, the left and right sensors are on the lighter surface, while the center one is on the dark line. This drives both motors in the forward direction.

The robot uses a differential steering mechanism, where the motor is turned off to turn the robot in that direction. Let's consider a scenario where the left sensor is on top of the dark line. In this case, the robot needs to make a correction to the left. We turn off the left motor, and so turn to the left until the left sensor is back on the lighter surface. You can learn more about line-following sensors at https://bit.ly/3LSF49u.

The line-following sensor is installed on the bottom side of the robot, as shown in the following figure:

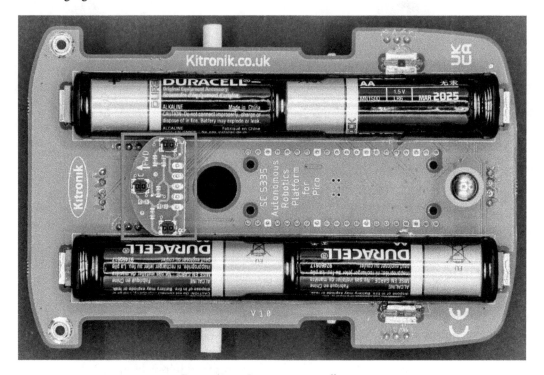

Figure 9.21 – Line sensor installation

Let's discuss the code needed to test the line-following sensor. It is available for download along with this chapter as line_following_sensor.py:

1. In the Thonny IDE, create a file called main.py and save it to your Pico.

2. The first step is to import the requisite modules. We are going to need the KitronikPicoRobotBuggy class:

```
from PicoAutonomousRobotics import KitronikPicoRobotBuggy
from time import sleep
```

3. The next step is to initialize an object belonging to the KitronikPicoRobotBuggy class:

```
buggy = KitronikPicoRobotBuggy()
```

4. Now, we should be able to retrieve the raw values of the line-following sensor as follows:

```
while True:
    print ("L" , buggy.getRawLFValue("l"),  " R", buggy.
getRawLFValue("r"),  " C",buggy.getRawLFValue("c"))
    sleep(1)
```

5. Putting it all together, we have the following:

```
from PicoAutonomousRobotics import KitronikPicoRobotBuggy
from time import sleep
buggy = KitronikPicoRobotBuggy()

while True:
    print ("L" , buggy.getRawLFValue("l"),  " R", buggy.
getRawLFValue("r"),  " C",buggy.getRawLFValue("c"))
    sleep(1)
```

6. When we execute the preceding code sample, we get the following output while trying to test the sensor output by placing the robot on a surface:

```
Shell ×
    L 14419    R 11922    C 4833
    L 14371    R 11794    C 4833
    L 14419    R 11890    C 4849
    L 14451    R 11938    C 4833
    L 14435    R 11858    C 4833
    L 14467    R 11938    C 4817
```

Figure 9.22 – Line-following sensor output

The preceding figure shows the raw sensor output of the left, right, and center sensors. The center sensor is showing a small value, while the left and right sensors are displaying a very high value. This means that the left and right sensors are on top of a darker surface, while the center one is on top of a lighter surface.

The raw sensor value can be used to steer the robot to follow a dark line on a lighter background. In the next section, we will discuss testing the robot.

Testing the robot

Now that we have tested the components of our robot, it is time to take it for a spin. The first step is to install the wheels of the robot (as shown in the following figure):

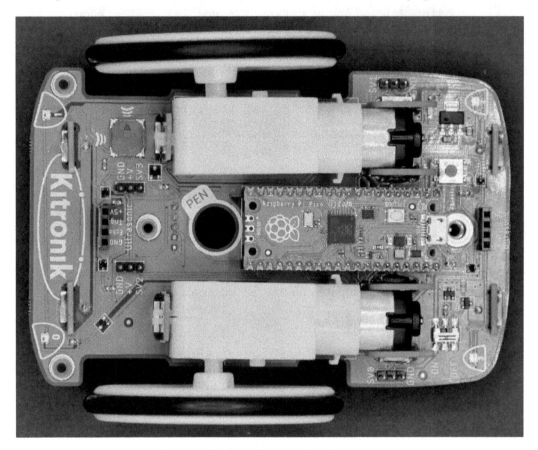

Figure 9.23 – Installing wheels on the robot

The code samples for the line-following and obstacle-avoidance robots are available for download along with this chapter as `line_following.py` and `obstacle_avoidance.py`. We tested the line-following robot using a mat from Kitronik (`https://bit.ly/3GAn0xM`), as shown in the following figure:

Figure 9.24 – Line-following robot on a map

Robotics contests

Where do we go from here? If building robots interests you, we recommend checking out contests organized by your local robotics club. There are other types of robots, such as maze-solving robots and sumo wrestling robots, for example. Some clubs even organize robotic sumo wrestling competitions. Our favorite contest is a Micromouse contest, where the robot is supposed to solve a maze and find its way home. The challenge is in finding the shortest way home. You can learn more about Micromouse contests at https://en.wikipedia.org/wiki/Micromouse. In the United States, the IEEE (IEEE is a technical professional association for engineers) is known to conduct regional Micromouse contests at student events.

Summary

In this chapter, we discussed building a robot using an off-the-shelf Kitronik Robotics kit. We discussed the individual components of the robot, including testing the LEDs, motor selection, and DC and servo motor control. We discussed drivers meant for driving the motors and interfacing the sensors. We also discussed testing the line-following robot and the obstacle-avoidance robot. This was followed by a review of how to take this project forward.

In the next chapter, we will discuss building TinyML applications on the Pico.

10
Designing TinyML Applications

In the previous chapters, our projects ranged from building a weather station to a line-following robot. In this chapter, we will discuss developing TinyML applications using the Pico. We will start by introducing TinyML and its potential uses. Then, we will discuss an example that involves image classification.

In this chapter, we will cover the following topics:

- Introducing TinyML
- Keyword recognition in audio samples
- Classifying images
- Developing edge devices

Technical requirements

The following hardware is required for this chapter:

- Raspberry Pi Pico (`https://bit.ly/3AJtoAf`): USD 4

- Arducam Camera Shield (`https://amzn.to/32KPqVN`): USD 25.99

- Adafruit PDM microphone breakout (`https://bit.ly/3tGCzRt`): USD 4.95

Optional hardware

We also recommend the following optional hardware:

- Arduino RP2040 Connect (`https://bit.ly/2YZClrY`): USD 24.50

- Arducam Pico4ML TinyML Dev Kit (`https://bit.ly/3DfKJTW`): USD 25.99

- Arduino Tiny Machine Learning Kit (`https://bit.ly/3M7m4nK`): USD49.99

The code samples in this chapter can be found in this book's GitHub repository at `https://github.com/PacktPublishing/Raspberry-Pi-Pico-DIY-Workshop/tree/main/chapter_10`.

Code in Action videos for this chapter can be viewed at `https://bit.ly/3yklorx`.

Now, let's introduce TinyML.

Introducing TinyML

What is TinyML? **TinyML** refers to **Tiny Machine Learning** and it is a nascent but growing field where **machine learning** (**ML**) tools are used on resource-constrained hardware, such as an RP2040 microcontroller, to interpret sensor data. The resource constraints refer to the limited memory and processing power available on a microcontroller compared to a server with enormous processing power and GPU. TinyML allows you to interpret data on a microcontroller powered by a coin cell. A device that can interpret sensor data using TinyML tools locally instead of having to upload the data to the cloud is called an **edge device**.

Let's illustrate this concept with an example. The following diagram shows the flow of data in a typical IoT application, where we have a device that is collecting data from various sensors and forwarding it to the cloud. The inference happens in the cloud and the server running in the cloud instructs the gateway to turn devices on/off:

Typical IoT Solution

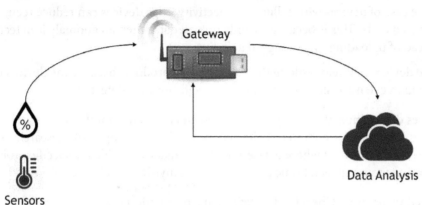

Figure 10.1 – Typical IoT solution

With an edge device, this flow of data is short-circuited where the inference happens locally at the gateway. There is still some data flowing to the cloud, but we only upload the events that were detected. These events include any anomalies such as leaks in the system, the presence or absence of objects in the system, predicting component failures from known noises, and more. This allows you to react to an anomaly instead of waiting for a response from the cloud:

An Edge Solution

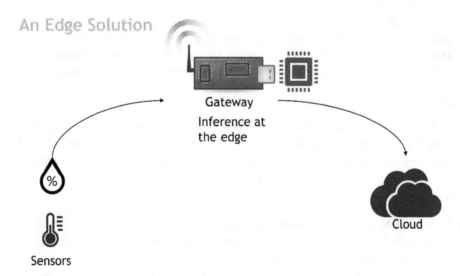

Figure 10.2 – Edge device-based solution

Edge devices come with certain advantages:

- In the case of devices with cellular connectivity, edge devices can reduce recurring overhead costs. This is because we only upload data when an anomaly is detected instead of uploading every single data point.

- Edge devices can help prolong the battery life of products because the devices can remain in extended sleep states unless a specific event is detected.

Edge devices detect anomalies using data that's been collected from the field. As a result, edge devices require some data to be collected for product development. Researchers are making inroads in this field where datasets are being made available for specific problems, such as bearing wear and tear. Some applications of TinyML include the following:

- **Visual inspection**: Checking for defects in a production line.

- **Predictive maintenance**: Using an accelerometer to detect unusual vibrations and impending failure.

- **Monitoring flora/fauna**: TinyML-enabled devices could be used to monitor the sounds of vehicles or hacksaws to prevent wildlife from being poached.

- **Image classification**: Tracking wildlife using trap cameras.

> **Smart Sensors**
>
> With the advent of powerful processors, semiconductor manufacturers are designing sensors with ML cores. This allows you to integrate an ultra-low-power sensor that can detect specific events. Here is a link to an inclinometer with an ML core: `https://www.st.com/en/mems-and-sensors/iis2iclx.html`.

Before we dive into a TinyML application example, let's briefly discuss how TinyML relates to the field of **artificial intelligence** (**AI**). TinyML is a small subset of ML, which, in turn, is a subset of the AI field. AI refers to enabling a machine to make decisions like a human being. ML refers to training a machine to learn from the collected data and make decisions or provide predictions on its own.

Within the field of ML, there are various toolsets and one of them is called deep learning. This is a technique that is modeled after the human brain, where information passes through a network of layers. As the information flows through different layers, the machine can "ingest" and process the information to make a prediction. This network of layers is called a neural network.

A neural network comprises of an interconnected system of nodes called **neurons**. These neurons are arranged into input, output, and one or more intermediate layers, as shown in the following diagram. The nodes of a neural network each have an associated weight assigned to them:

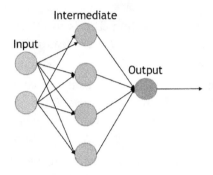

Figure 10.3 – Neural network model

Neural networks typically require vast amounts of processing power and, as described earlier, they are usually run on a server. TinyML's tools allow you to run a modified neural network on an RP2040 microcontroller.

Let's consider an image classification example where we are trying to determine whether an image belongs to a particular class. In this case, the neural network accepts an image as input and provides the probability of whether an image belongs to a particular class.

You can learn more about neural networks at `https://news.mit.edu/2017/explained-neural-networks-deep-learning-0414`.

The following diagram shows the various stages involved in deploying a neural network onto a microcontroller:

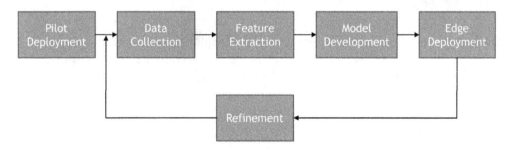

Figure 10.4 – Stages involved in deploying a neural network onto a microcontroller

Let's look at these stages in more detail:

1. The first step is data collection. This is accomplished by collecting data from devices currently being used in the application or deploying a pilot run of devices that are used to collect data.

2. The next step is called feature extraction. This is also the step where we determine the features/anomalies. If a human can detect a feature/anomaly by looking at the data, a neural network can be trained to do the same.

3. The next step is to train the neural network using the collected data. The collected data is split into validation and training datasets. After training, the neural network's performance is validated using the other dataset. The parameters of the neural network are fine-tuned over several training cycles.

4. Once the neural network is ready, we convert it into a format that is suitable for a microcontroller. This process is called **quantization**. This is because a neural network's weights consist of floating-point numbers. During quantization, the weights are converted into integers. This speeds up the performance of the neural network running on the microcontroller.

5. Once the model has been quantized, it is time to deploy it on a microcontroller and measure its performance.

6. The model is further fine-tuned based on the performance measurements. This is because there is a drop in performance due to quantization.

Online Course for TinyML

If you are interested in pursuing a career in this field, we recommend the following course from edX and Harvard: `https://www.edx.org/professional-certificate/harvardx-tiny-machine-learning`.

In this chapter, we will demonstrate two examples of deploying a neural network for image classification and keyword recognition. In the next section, we will briefly discuss a hardware development kit meant for TinyML applications.

Introducing the Pico4ML

The Pico4ML is an RP2040 microcontroller development board from ArduCam:

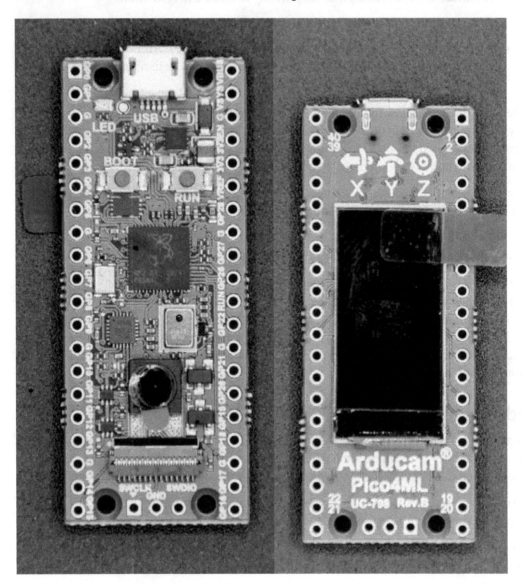

Figure 10.5 – A Pico4ML development board

The board is equipped with the following sensors:

- A HiMax camera module that supports up to 320 x 240 resolution.

- A microphone with **pulse density modulation** (PDM) output.

- An ICM-20948 9-axis **inertial measurement unit** (IMU). An IMU is a combination of a gyroscope, an accelerometer, and a compass. It can be used in gesture recognition applications.

These three sensors allow the development board to be used in image classification, keyword spotting, and gesture recognition applications, respectively. The maker of the development board has provided examples to help users get started with this board. They can be found at `https://github.com/ArduCAM/RPI-Pico-Cam`.

The examples discussed in this section can be executed using this development board. In the next section, we will introduce image classification applications.

Keyword recognition in audio samples

In this section, we will discuss keyword recognition in an audio sample using the Pico. This is similar to the voice assistant devices that are designed and marketed by tech giants such as Amazon Echo, Google Home, and others that respond to an activation keyword such as "Hey Alexa." We will be making use of Edge Impulse (`https://edgeimpulse.com`) to work on this example. Sign up for a developer account on Edge Impulse to work along with this example.

Edge Impulse

Edge Impulse is a platform that allows you to develop TinyML applications using their platform and deploy them on a microcontroller. Edge Impulse is an interactive tool that allows you to develop applications without prior knowledge of machine learning.

If you are interested in becoming a certified developer, we recommend these courses from Edge Impulse on Coursera: `https://www.coursera.org/edgeimpulse`.

We worked on this example while using the following tutorial as a guide: `https://docs.edgeimpulse.com/docs/tutorials/audio-classification`.

There's also our project, which you can clone and follow along with. It can be found at `https://studio.edgeimpulse.com/public/48695/latest`.

The **Clone this project** button can be found in the top-right corner, as shown in the following screenshot. You will need a developer account to clone the project. It will take a few minutes for the project to be cloned into your account:

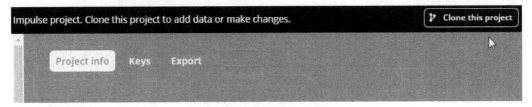

Figure 10.6 – Cloning the project

Once completed, it should display the following message:

Clone succeeded

You're now ready to build your next embedded Machine Learning project!

Figure 10.7 – Clone succeeded

Let's discuss the steps involved in developing the keyword recognition example using Edge Impulse:

1. Upon launching the project, your dashboard should look something like this:

Creating your first impulse (100% complete)

Acquire data

Every Machine Learning project starts with data. You can capture data from a development board or your phone, or import data you already collected.

LET'S COLLECT SOME DATA

Design an impulse

Teach the model to interpret previously unseen data, based on historical data. Use this to categorize new data, or to find anomalies in sensor readings.

🏃 GETTING STARTED: CONTINUOUS MOTION RECOGNITION

🎤 GETTING STARTED: RESPONDING TO YOUR VOICE

📷 GETTING STARTED: ADDING SIGHT TO YOUR SENSORS

Deploy

Package the complete impulse up, from signal processing code to trained model, and deploy it on your device. This ensures that the impulse runs with low latency and without requiring a network connection.

</> DEPLOY YOUR MODEL

Figure 10.8 – Project dashboard for the keyword recognition example

2. The first step in the development process is data acquisition (according to *Figure 10.3*). Click on **LET'S COLLECT SOME DATA**, as shown in the preceding screenshot.

3. You will be presented with various options to collect audio data for our application. The easiest way to collect data is by using a smartphone or microphone:

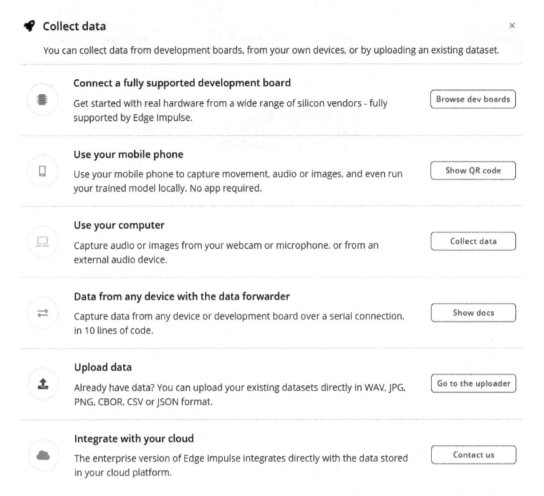

Figure 10.9 – Options to collect data

4. For example, when you click on **Use your computer**, you should be able to collect and label audio samples from your desktop:

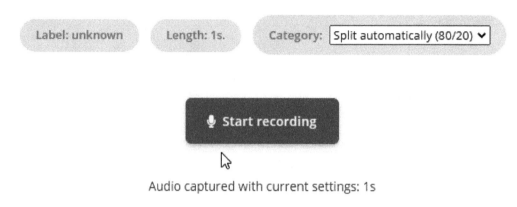

Figure 10.10 – Collecting audio samples using a computer

5. For this project, we used the audio dataset available at `https://cdn.edgeimpulse.com/datasets/keywords2.zip`. The dataset contains audio samples belonging to four classes, namely the words **yes**, **no**, **unknown samples of other words**, and **background noise**.

6. Now that we have uploaded the audio samples to our project and labeled them, we need to label the samples to identify the class of each audio sample. The following figure shows the audio samples and their corresponding labels:

SAMPLE NAME	LABEL	ADDED	LENGTH	
yes.ffb86d3c_nohash_0.wav....	yes	Today, 15:00:30	1s	⋮
yes.ff63ab0b_nohash_0.wav....	yes	Today, 15:00:30	1s	⋮

Figure 10.11 – Audio datasets uploaded to Edge Impulse

7. The next step is to create an impulse. An impulse takes the audio samples, processes the sample, and extracts the features to determine whether the audio sample belongs to one of the four classes in our audio dataset. An impulse can be created from the **Create impulse** tab on the left. Then, click on the **Add a Time series data** block. You also need to add an **Audio (MFCC)** processing block and a **Classification (Keras)** block before saving the impulse:

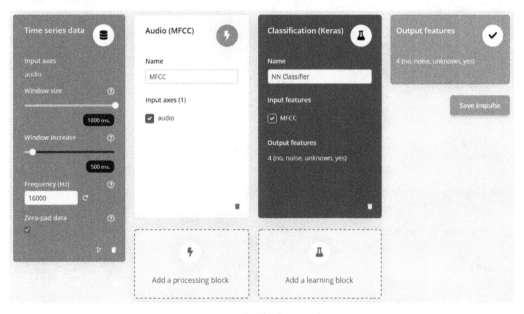

Figure 10.12 – Impulse for keyword recognition

8. Now, let's go to the **MFCC** tab on the left. In this project, keyword recognition can be accomplished using a spectrogram. A spectrogram is a visualization of all the frequencies in an audio sample. An MFCC spectrogram highlights the frequencies that are common in human speech. For now, we will leave the default parameters as is and save them:

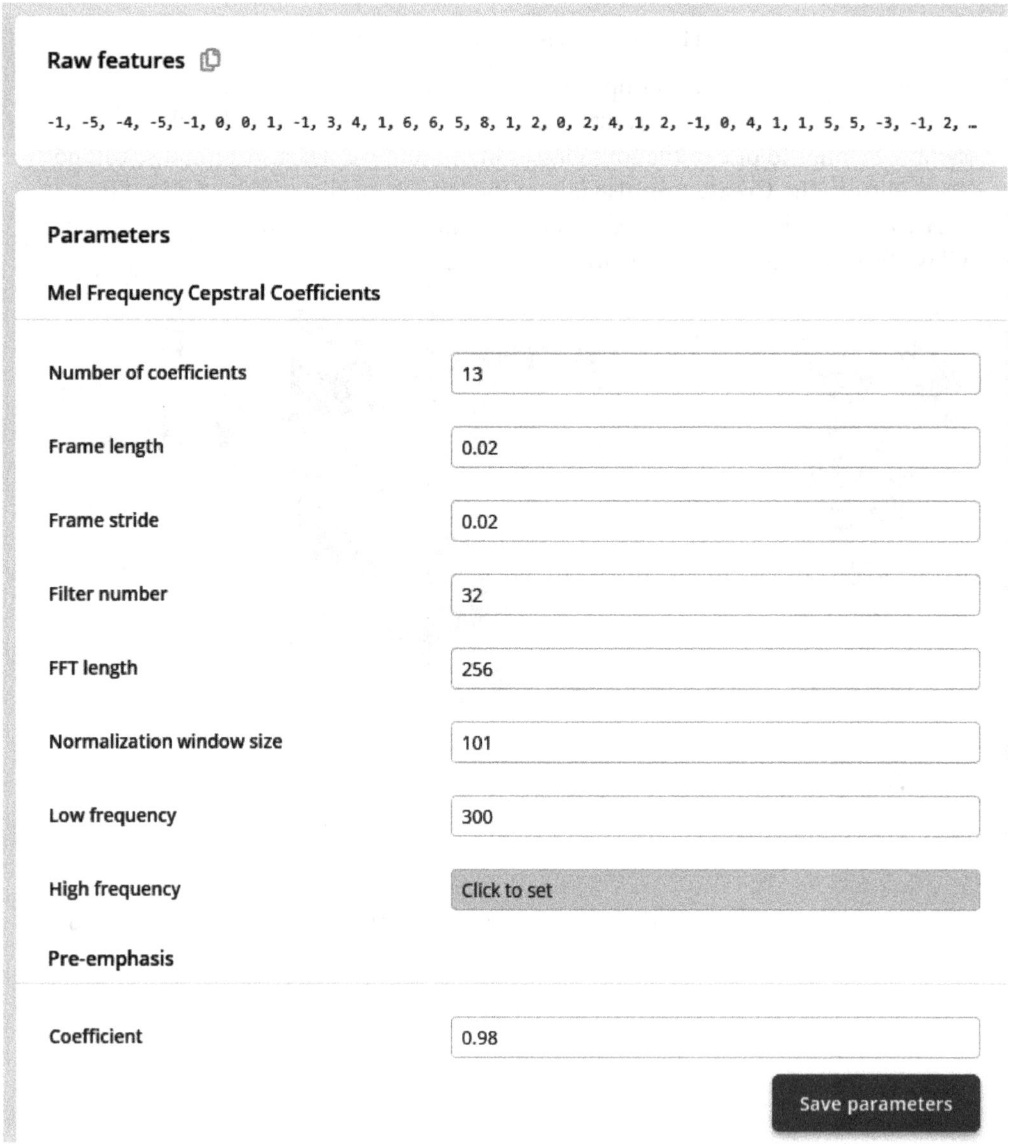

Raw features 📋

-1, -5, -4, -5, -1, 0, 0, 1, -1, 3, 4, 1, 6, 6, 5, 8, 1, 2, 0, 2, 4, 1, 2, -1, 0, 4, 1, 1, 5, 5, -3, -1, 2, …

Parameters

Mel Frequency Cepstral Coefficients

Number of coefficients	13
Frame length	0.02
Frame stride	0.02
Filter number	32
FFT length	256
Normalization window size	101
Low frequency	300
High frequency	Click to set

Pre-emphasis

Coefficient	0.98

Save parameters

Figure 10.13 – Save parameters

9. Additionally, we can see the performance of the MFCC processing block on an RP2040 microcontroller:

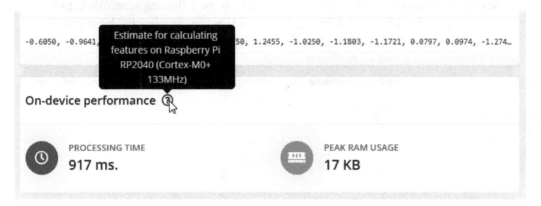

Figure 10.14 – On-device performance for the RP2040 microcontroller

10. Now, go to **Generate features** and click the **Generate features** button, as shown in the following screenshot:

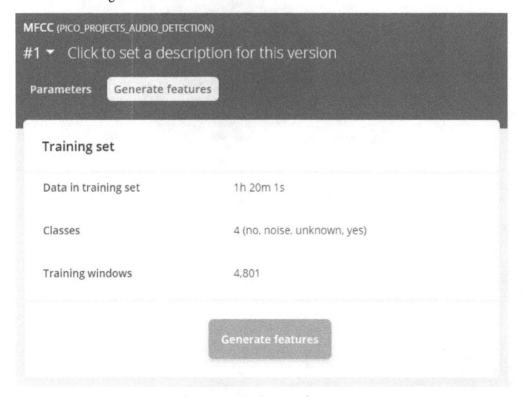

Figure 10.15 – Generate features

11. Once the features have been generated, you can visualize the grouping of your dataset, as shown in the following screenshot. This helps determine whether you can run machine learning algorithms on your data. In the following screenshot, each class is represented by a color. You can observe that the samples that belong to the four classes belong to four different clusters. This gives us confidence in being able to recognize the *yes* or *no* keywords:

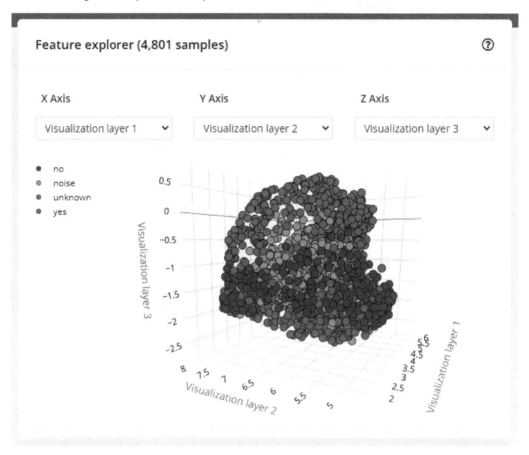

Figure 10.16 – Feature explorer

12. Next, let's look at the neural network classifier, where we train our model using the default parameters and the current dataset:

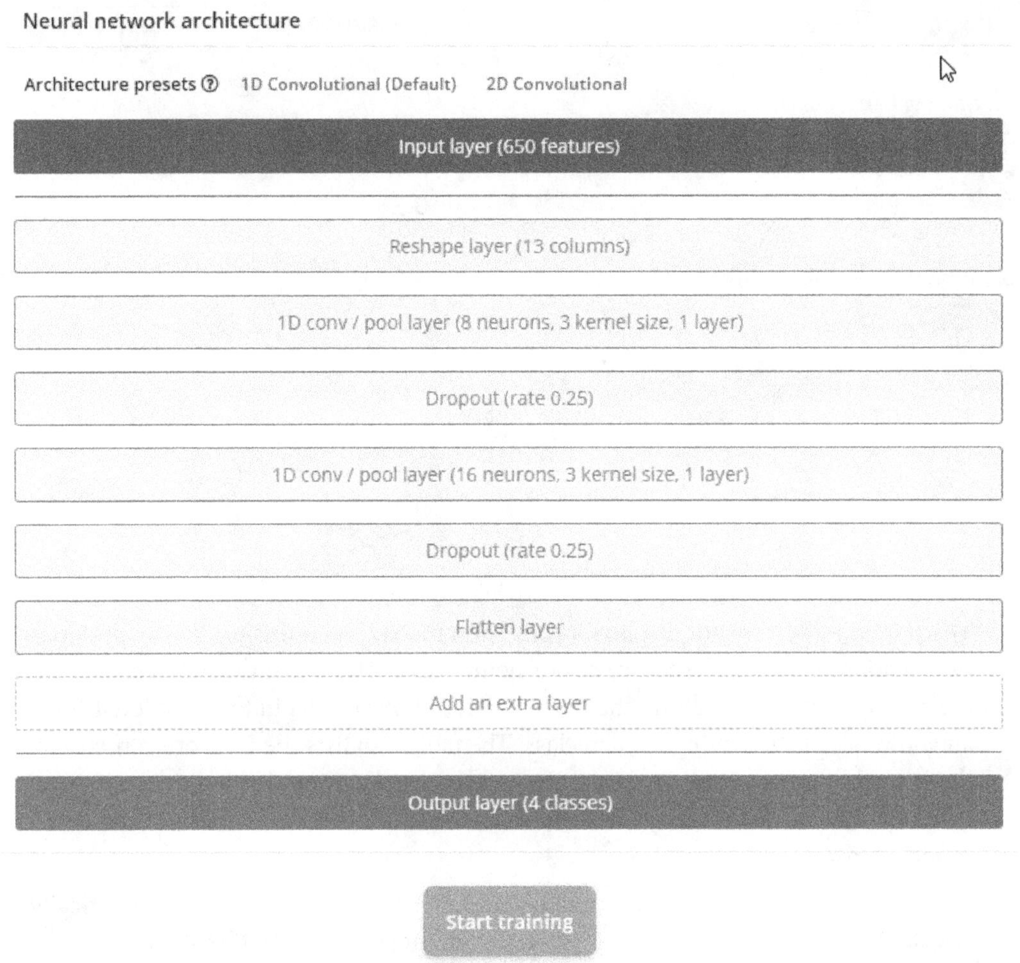

Neural network architecture

Architecture presets ⑦   1D Convolutional (Default)   2D Convolutional

Input layer (650 features)

Reshape layer (13 columns)

1D conv / pool layer (8 neurons, 3 kernel size, 1 layer)

Dropout (rate 0.25)

1D conv / pool layer (16 neurons, 3 kernel size, 1 layer)

Dropout (rate 0.25)

Flatten layer

Add an extra layer

Output layer (4 classes)

Start training

Figure 10.17 – Start training

13. The training can take a while but once it's complete, we can observe the training output. The training output should look similar to the following:

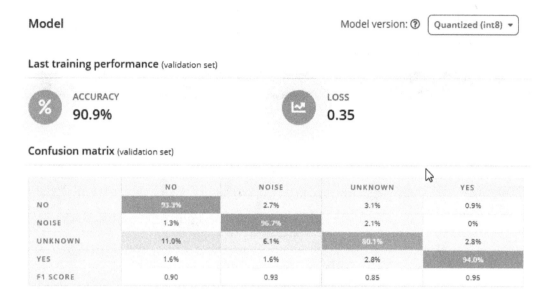

Figure 10.18 – Training output

The preceding screenshot displays a confusion matrix. A confusion matrix provides information on the performance of our neural network against the dataset. The table lists the statistics behind the performance of our model in terms of identifying an audio sample belonging to each class. The table also lists the F1 score, a metric that's used to measure the performance of our model without having to run it on a Pico. The F1 score can be anywhere between 0 and 1. A higher score indicates better model performance.

14. Next, let's look at the **EON Tuner** tab on the left-hand side. The tuner automatically selects the best-performing architecture, depending on the target hardware application. The target hardware is a Raspberry Pi RP2040 for this project. We can set it by clicking on the gear icon in the **Target** tab. The next step is to tune the model for the Pico. This can take a while:

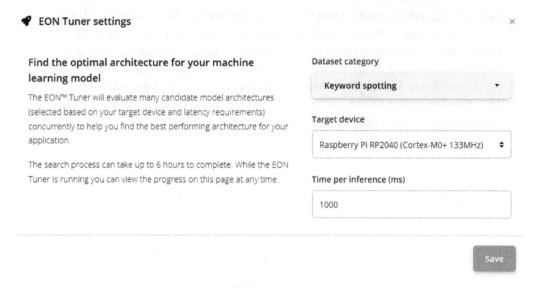

Figure 10.19 – Setting the target to Raspberry Pi RP2040

15. Once the target has been set, we can start the *EON tuner*, which selects the best architecture for our application. This can take a while:

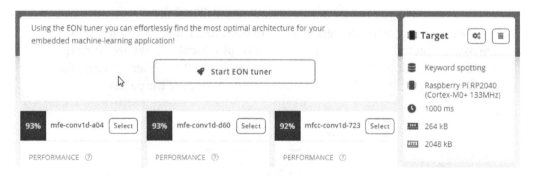

Figure 10.20 – Start EON tuner

16. Now, we can test the model from the **Model Testing** tab. When we uploaded our dataset to the Edge Impulse platform, the data was split in an 80:20 ratio. The data was arbitrarily split, where 80% of the uploaded samples are used for training and the remaining 20% are used for evaluating our model's accuracy. Splitting the dataset helps determine if the model has been overtrained on a dataset. This phenomenon is known as overfitting. The following screenshot shows the performance of our model against the test dataset:

Model testing results

% ACCURACY
84.45%

	NO	NOISE	UNKNOWN	YES	UNCERTAIN
NO	90.8%	0.3%	3.0%	0.3%	5.6%
NOISE	0.3%	91.1%	3.6%	0.3%	4.6%
UNKNOWN	8.4%	3.9%	65.9%	2.3%	19.5%
YES	1.6%	0.3%	1.3%	90.3%	6.5%
F1 SCORE	0.90	0.93	0.76	0.93	

Figure 10.21 – Model performance against the test dataset

17. Since we are happy with the performance of our model, it is time to test it by deploying it on the Pico. To do so, go to the **Deployment** tab. We are going to download an Arduino library to run the model on the Pico. After selecting the Arduino library, click on the **Build** button at the bottom of the page:

Create library

Turn your impulse into optimized source code that you can run on any device.

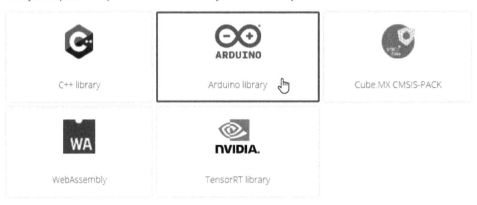

Figure 10.22 – Downloading the Arduino library

> **Programming the Pico Using an Arduino**
>
> If you are not familiar with the Arduino development environment, we recommend reading through *Chapter 12, Best Practices for Working with the Pico*. Ensure that the necessary board packages have been installed for the RP2040 Connect and the Pico.

18. Once the library has finished building, a ZIP file will become available for you to download from the Edge Impulse platform. The downloaded ZIP file will contain the libraries and code samples needed to run the model on a Raspberry Pi Pico.

19. In the Arduino IDE, go to **Sketch | Include Library | Add .ZIP Library** and include the downloaded ZIP file:

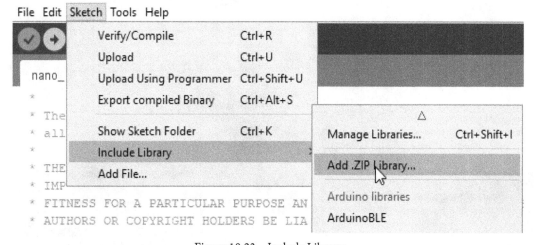

Figure 10.23 – Include Library

20. Now, the code sample will be available at **File | Examples | Pico_Projects_Audio_ Detection_inferencing | nano_ble33_sense_microphone**. `Pico_projects_ Audio_Detection_inferencing` is the name of our Edge Impulse project:

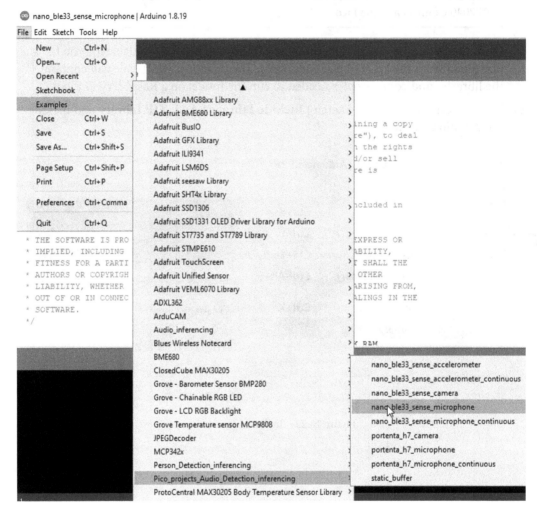

Figure 10.24 – The code sample's location

21. Now, we need to verify whether our code sample compiles successfully for an RP2040 microcontroller. We can test this by setting it to Arduino RP2040 Connect. This is because the code sample is designed to use the PDM microphone on an Arduino Nano board. We are ensuring that our code compiles successfully before we make modifications for the Pico:

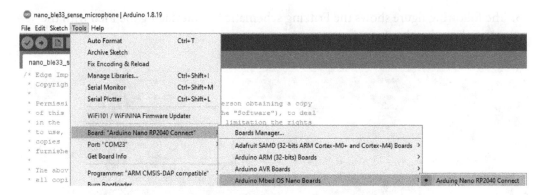

Figure 10.25 – Setting the target to Arduino Nano RP2040 Connect

22. You can verify that the code compiles by clicking on the **Verify** button (we are assuming that you are either familiar with the Arduino IDE or familiarized yourself by going through *Chapter 12, Best Practices for Working with the Pico*).

23. If you have an Arduino RP2040 Connect, you can upload the compiled binary by clicking on the **Upload** button. Now, you are ready to test your code by launching the Serial Monitor in the Arduino IDE. This is because of the PDM microphone that's available on the Arduino RP2040 Connect. You can skip to *Step 28* if you wish to test the application. The code sample for the Arduino RP2040 Connect is called `arduino_rp2040_connect_sense_microphone.ino` and can be downloaded from this book's GitHub repository.

24. If you are working with a Pico, then you need to wire up the PDM microphone breakout to the Pico as follows, where the left-hand side of the arrow refers to a pin on the microphone breakout, while the right-hand side of the arrow refers to a pin on the Pico:

- 3.3V → 3.3V
- GND pins tied together
- SEL → 3.3V
- CLK → GP3
- DAT → GP2

25. The following figure shows the Fritzing schematic for interfacing the Pico with the microphone breakout:

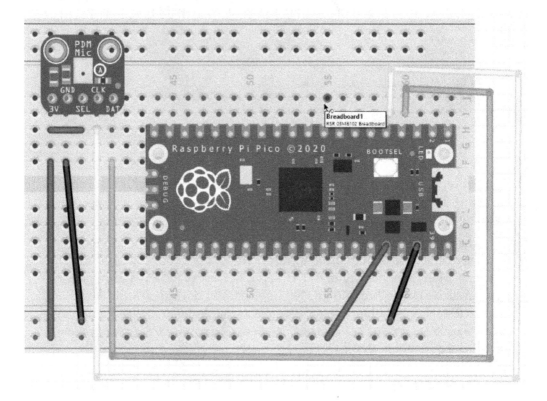

Figure 10.26 – Fritzing schematic for the PDM microphone breakout

26. We need to change the target to the Pico. We also need to make some minor modifications to our code sample (two macro definitions need to be included). Once the code has been compiled and uploaded to the Pico, it is time to take our project for a spin. The code sample for this is called `Pico_microphone.ino` and can be found in this book's GitHub repository.

27. Upon launching the serial port, you should see the following output in your serial port monitor, where the model running on the microcontroller is providing a prediction score for the collected audio sample. In the following screenshot, we can see that the model can predict that the collected audio sample is noise with 99% confidence:

```
    no: 0.00391
    noise: 0.91016
    unknown: 0.08203
    yes: 0.00391
Starting inferencing in 2 seconds...
Recording...
Recording done
Predictions (DSP: 1477 ms., Classification: 25 ms., Anomaly: 0 ms.):
    no: 0.01562
    noise: 0.85547
    unknown: 0.12500
    yes: 0.00391
Starting inferencing in 2 seconds...
Recording...
Recording done
Predictions (DSP: 1476 ms., Classification: 25 ms., Anomaly: 0 ms.):
    no: 0.00000
    noise: 0.99219
    unknown: 0.00781
    yes: 0.00000
Starting inferencing in 2 seconds...
```

☑ Autoscroll ☐ Show timestamp

Figure 10.27 – Noise in the audio sample

28. When we utter the word *yes*, the model can predict with 99% percent confidence that the audio sample contained the word *yes*:

```
    no: 0.80859
    noise: 0.01172
    unknown: 0.02734
    yes: 0.15234
Starting inferencing in 2 seconds...
Recording...
Recording done
Predictions (DSP: 1490 ms., Classification: 25 ms., Anomaly: 0 ms.):
    no: 0.00781
    noise: 0.98047
    unknown: 0.00781
    yes: 0.00391
Starting inferencing in 2 seconds...
Recording...
Recording done
Predictions (DSP: 1485 ms., Classification: 25 ms., Anomaly: 0 ms.):
    no: 0.00000
    noise: 0.00000
    unknown: 0.00000
    yes: 0.99609
Starting inferencing in 2 seconds...
```

Figure 10.28 – Audio sample containing yes (99% confidence)

29. When we utter the word *no*, the model can predict with 78% confidence that the audio sample contains the word *no*:

```
    unknown: 0.02734
    yes: 0.00000
Starting inferencing in 2 seconds...
Recording...
Recording done
Predictions (DSP: 1483 ms., Classification: 25 ms., Anomaly: 0 ms.):
    no: 0.00000
    noise: 0.00000
    unknown: 0.00000
    yes: 0.99609
Starting inferencing in 2 seconds...
Recording...
Recording done
Predictions (DSP: 1482 ms., Classification: 25 ms., Anomaly: 0 ms.):
    no: 0.78906
    noise: 0.00781
    unknown: 0.16797
    yes: 0.03125
Starting inferencing in 2 seconds...
Recording...
Recording done
```

☑ Autoscroll   ☐ Show timestamp

Figure 10.29 – Audio sample containing no (78% confidence)

With that, we have successfully built a TinyML application that can detect a keyword from an audio sample. In the next section, we are going to discuss a pre-built example that detects whether a person is present in a frame that's been captured by a camera.

Classifying images

In this section, we will discuss an image classification example using the Pico. We will use a binary that has been already compiled and can be downloaded to determine whether a person is present in the frame that's been captured by a camera. Let's get started:

1. First, download the compiled binary from `https://github.com/ArduCAM/RPI-Pico-Cam/blob/master/tflmicro/bin/person_detection_screen_int8.uf2`.

2. The next step is to put the Pico in bootloader mode to load the binary. We recommend checking out *Chapter 1*, *Getting Started with the Raspberry Pi Pico*, if you are not familiar with the process.

3. Copy over the downloaded binary to the Pico.

4. Now, we need to interface the camera module to the Pico as follows, where the left-hand side of the arrow refers to a pin on the camera module and the right-hand side of the arrow refers to a pin on the Pico:

   - CS → GP5

   - MOSI → GP3

   - MISO → GP4

   - SCK → GP2

   - GND pins tied together

   - VCC → 3.3V

   - SDA → GP8

   - SCL → GP9

5. Now, it is time to test our application using the Pico. We used a Pico4ML developer board to test the binary. This is because the board comes with an onboard camera module. We pointed the camera at a bobblehead and we observed the following output:

Figure 10.30 – Pico4ML pointed at a bobblehead

6. The Pico can predict with 93.8% confidence that the captured frame contains a person. This is because the model tries to detect facial features in the frame to detect whether a person is present in the frame.

7. The Pico also outputs the inference statistics to the serial port, as shown in the following screenshot:

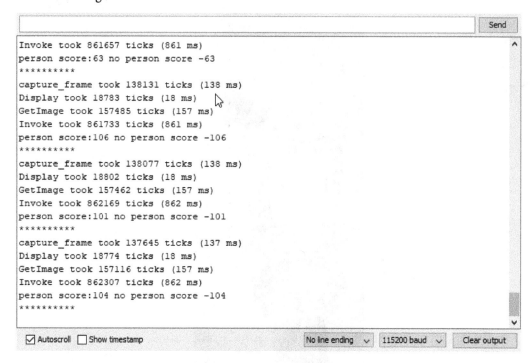

```
                                                                              Send

Invoke took 861657 ticks (861 ms)
person score:63 no person score -63
**********
capture_frame took 138131 ticks (138 ms)
Display took 18783 ticks (18 ms)
GetImage took 157485 ticks (157 ms)
Invoke took 861733 ticks (861 ms)
person score:106 no person score -106
**********
capture_frame took 138077 ticks (138 ms)
Display took 18802 ticks (18 ms)
GetImage took 157462 ticks (157 ms)
Invoke took 862169 ticks (862 ms)
person score:101 no person score -101
**********
capture_frame took 137645 ticks (137 ms)
Display took 18774 ticks (18 ms)
GetImage took 157116 ticks (157 ms)
Invoke took 862307 ticks (862 ms)
person score:104 no person score -104
**********

☑ Autoscroll  ☐ Show timestamp            No line ending ⌄   115200 baud ⌄   Clear output
```

Figure 10.31 – Serial port output from Pico

That concludes our image classification example of using the Pico.

Developing edge devices

There are some factors to consider before you start developing a product that makes use of TinyML, as follows:

- The datasets that are available for developing your product. The best place to start is by making use of existing datasets or making use of data from an existing application.

- Product development cycles involving TinyML require a lot of trial and error to refine the parameters.

- TinyML applications are suitable for tapping new revenue streams and improving productivity.

- You also need to account for retraining your model from time to time to account for problems identified in the system.

Now that we have discussed some of the factors to consider before developing a product, let's summarize this chapter.

Summary

In this chapter, we introduced the topic of TinyML, its applications, and getting started with TinyML development. We also discussed a keyword recognition example using Edge Impulse and discussed testing it with a Pico. We also discussed an image classification example where we used a compiled binary to detect whether a person is present in the frame. We pointed a camera at a bobblehead to demonstrate this.

Finally, we wrapped up this chapter by discussing some factors to consider while developing an edge device.

In the next chapter, we will discuss developing a product around the Pico.

11
Let's Build a Product!

In this chapter, we will discuss building a product using the Raspberry Pi Pico. We will design a carrier **Printed Circuit Board (PCB)** for the Pico where we will interface a cellular module to send messages to the cloud. We will also discuss how to build a Pico phone, as shown in the following figure. In this chapter, we will specifically discuss the product development process that's involved in designing a PCB.

Figure 11.1 – Pico phone

In this chapter, we will cover the following topics:

- Understanding the Pico phone
- Capturing the requirements
- Designing a PCB
- Bringing up the board
- Taking the project forward

Technical requirements

The software and hardware files for this chapter can be found at `https://bit.ly/35088w4`.

The following hardware is required for this chapter:

- Raspberry Pi Pico (`https://bit.ly/3AJtoAf`): USD 4.
- Blues Wireless Notecard (`https://bit.ly/3wnFbWc`): USD 65.
- Cellular Antenna (`https://bit.ly/3N91Sn1`): USD 11.50.
- Qwiic Keypad (`https://bit.ly/3JwIJJd`): USD 11.50.
- Fabricated PCB: The prices vary based on the manufacturer.
- PCB components: The prices depend on the components. The **Bill of Materials (BoM)** is available in this book's GitHub repository (shared previously).
- Soldering supplies, including a reflow oven.

Code in Action videos for this chapter can be viewed at `https://bit.ly/38Th2gy`.

> **Installing CircuitPython**
>
> We are assuming that you have installed CircuitPython on your Pico. If you are not familiar with the installation process, we recommend following the installation process provided in *Chapter 1*, *Getting Started with the Raspberry Pi Pico*.

In the next section, we will briefly discuss the Pico phone and its origins.

Understanding the Pico phone

We built the Pico phone in 2021 and published it as a Hackster project (`https://bit.ly/3pY7J5e`). The project originated when we wanted to evaluate a cellular module. We chose this project to describe the product development process because it was one of 21 featured projects of 2021 (`https://bit.ly/3CXh3ec`) and it was also featured in the March 2022 edition of the Hackspace magazine (`https://bit.ly/3CYAJ11`), which was published by the Raspberry Pi foundation. The magazine is free to download.

> **Hackster Article versus This Chapter**
>
> While the Hackster project discusses texting a joke to a phone number, we will discuss the product development aspects of the project. Specifically, we will discuss capturing the requirements, building the prototype, and more.

In the next section, we are going to discuss capturing the requirements for the product.

Capturing the requirements

In this section, we will discuss capturing the requirements for our product. Before we start developing a product, we need to capture the requirements for our product. We also need to understand whether it is feasible to make a product.

To determine this feasibility, we need to understand our product's purpose. Let's define some requirements for our product:

- The product needs to have some form of network connectivity to connect and upload data to the cloud.

- The product needs to have a keypad to capture user inputs.

- The product needs to use a Raspberry Pi Pico. This will allow you to use the RP2040 microcontroller in future revisions. The first version would make use of the Pico, while future revisions would use the RP2040 microcontroller integrated directly onto the PCB.

- The device can be powered using a USB cable (by using the Pico's USB connector) or using a DC power jack.

In the real world, the product would have more requirements. We are keeping it simple in this chapter.

Selecting components

Now that we have understood the requirements, let's discuss component selection for
the product:

- We are going to be using the Raspberry Pi Pico. It is important to understand the
  peripherals that are available on the Pico to identify the network connectivity
  options. Usually, an exhaustive search is needed to find a microcontroller to capture
  all the requirements of our product. Some of the options to consider include unit
  price, footprint, and availability.

- Another reason to choose the Pico for our product is because of the RP2040
  microcontroller. When we migrate our design from the Pico to the RP2040
  microcontroller, it helps keep the BoM costs low since the RP2040 microcontroller
  costs only USD 1. We would still need other components to build a minimal
  circuit. We will discuss migrating our product from the Pico into an RP2040
  microcontroller-based design in the *Taking the project forward* section.

- If you are not familiar with the component search and selection process, we
  recommend the weekly segment from Adafruit Industries called the *Great Search*.
  For example, here is a segment where they demonstrate selecting a microcontroller
  for a product: `https://bit.ly/3rfDYg3`.

> **Checking for Product Availability**
>
> During the component selection stage, it is important to ensure that the
> component in question is available in the quantities you need during the
> manufacturing stage. This is especially true during the semiconductor shortage
> due to the ongoing pandemic. It would be ideal to select components with a
> drop-in replacement. This would allow you to replace the component in the
> event of a shortage.

- We need to identify connectivity options for our product. The different connectivity
  options include Wi-Fi, Bluetooth, and cellular. We chose cellular connectivity for
  the following reasons:

  - We want our product to be provisioned and used with relative ease. Devices with
    cellular connectivity can be provisioned before being shipped to the customer.

  - If we use Wi-Fi, we will need to provide detailed instructions for the customer
    to set up the device. We will also need to ensure that the device is installed at
    a location that has decent Wi-Fi connectivity.

- Cellular networks have certain limitations, such as network coverage, recurring costs, and more. We are going to assume that our product is going to be used in areas where cellular connectivity is available.

• For cellular connectivity, we chose to use the Notecard (shown in the following figure) from `https://blues.io/`:

Figure 11.2 – Notecard from Blues.io

We chose it for the following reasons:

- It comes with 10 years and 500 MB of free cellular data. This allows you to design a product that doesn't have a recurring cost.

- It can be controlled via the I2C interface or the UART interface. The Pico has both interfaces.

- It comes with Python libraries that allow you to interface with the Pico using CircuitPython.

- We need a keypad to capture user inputs. We chose SparkFun's Qwiic Keypad (shown in the following figure) for our product because it helps eliminate the need for code to implement keypad scanning. It is offloaded to the keypad, which acts as an I2C peripheral.

Figure 11.3 – SparkFun Qwiic Keypad (image source: SparkFun; License: CC by 2.0)

- We will also make use of a Qwiic/Stemma connector in our design that would enable interfacing any I2C peripheral to our product.

Now that we have identified the major components of our product, it is time to build a proof-of-concept for the product.

Building a proof-of-concept

In this section, we are going to discuss building a proof-of-concept for the product. This includes verifying that the major components are working.

Prototype versus Proof-of-Concept

A proof-of-concept refers to proving the functional requirements of the product. It may not look like the final product and it is meant to identify components that meet the requirements of our product. A prototype refers to something that looks like a final product. A prototype might not be on the same scale as the final product but a prototype is used to fine-tune the features of our product.

For this product, we need to test the following features:

- Verify the Notecard interface and cellular connectivity
- Verify the keypad and input capture

To verify the proof of concept, we need to identify and install the requisite libraries. We are going to do this in the next section.

Installing the requisite libraries

While we're building the proof-of-concept, we also need to investigate the software that's required to build our product. This includes investigating whether libraries are available to interface the peripherals of the product. This helps determine the total time and resource investment for writing any required libraries. In this case, we managed to locate the libraries for both the cellular module and the keypad. Let's discuss how to install the libraries for both peripherals.

Cellular module

The cellular module's library is available at `https://bit.ly/37DlpeO`. Extract the contents of the latest version and copy them over from the `notecard` folder to the `lib` folder of your Raspberry Pi Pico.

Keypad

The Qwiic keypad library is available at `https://bit.ly/3L3EHZu`. Extract the contents of the latest version and copy over the `sparkfun_qwiickeypad.mpy` binary from the `lib` folder to the `lib` folder of your Raspberry Pi Pico.

Now that we have installed the requisite libraries, we will verify their functionality separately. In the next section, we will verify the functionality of the Notecard.

Testing the Notecard

In this section, we will interface the Notecard to the Raspberry Pi Pico and verify its functionality. The Notecard is a System-on-Module that comes in an M.2 form factor that needs a carrier board (shown in the following figure).

We will be using the following carrier board to test the Notecard interface (`https://bit.ly/34Yvnq`).

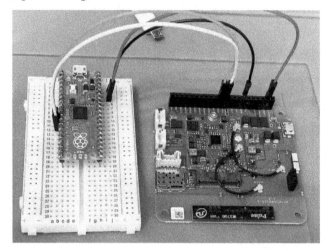

Figure 11.4 – Notecard interfaced to Pico (image source: hackster.io; License: MIT)

The Notecard is interfaced to the Raspberry Pi Pico as follows (as shown in the following figure):

- GP9 → SCL
- GP8 → SDA
- 3.3 V → VIN
- The GND pins tied together

The Fritzing schematic is as follows:

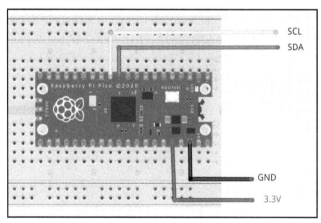

Figure 11.5 – The Fritzing schematic for interfacing using the I2C interface

Before we start testing the Notecard, we need to create a project on `https://notehub.io`. Creating a project allows us to upload data from the Notecard to the cloud. We recommend following the instructions at `https://bit.ly/3D1qpFZ` to create a new project. Once a new project has been created, make a note of the **product UID** since it is necessary for testing the Notecard.

Let's take a quick look at the code that's needed to verify the Notecard. The code sample we'll discuss in this section can be downloaded from this chapter's GitHub repository as `code_notecardtest.py`:

1. First, let's import the requisite modules. This includes the `board`, `busio`, and `notecard` modules:

```
import board
import busio
import notecard
```

2. Then, we must initialize the I2C interface, followed by the Notecard interface:

```
i2c = busio.I2C(board.GP9, board.GP8)
card = notecard.OpenI2C(i2c, 0, 0, debug=True)
```

3. In the previous code snippet, we initialized the Notecard with `debug=True`. This prints out all outgoing/incoming information.

4. Now, we need to include the Product UID that we retrieved from `https://notehub.io/sign-in?flow=bccb016e-6b5f-42e8-8f61-b9b4d463598b` earlier:

```
productUID = "com.example.name:project_name"
```

5. Let's set the product UID and connect the Notecard to the Notehub for the first time:

```
req = {"req": "hub.set"}
req["product"] = productUID
req["mode"] = "periodic"
req["inbound"] = 120
req["outbound"] = 60
req["sync"] = True
rsp = card.Transaction(req)

while True:
    pass
```

6. Putting it altogether, we get the following:

```
import board
import busio
import notecard
productUID = "com.example.name:project_name"
# Create bus object using our board's I2C port
i2c = busio.I2C(board.GP17, board.GP16)

# Create relay object
card = notecard.OpenI2C(i2c, 0, 0, debug=True)

req = {"req": "hub.set"}
req["product"] = productUID
req["mode"] = "periodic"
req["inbound"] = 120
req["outbound"] = 60
req["sync"] = True
rsp = card.Transaction(req)

while True:
    pass
```

7. When we save the preceding code sample as code.py, we should see the following output:

```
Auto-reload is on. Simply save files over USB to run them or enter REPL to disable.

code.py output:
{"inbound": 120, "outbound": 60, "mode": "periodic", "req": "hub.set", "product": '        , "sync": true}
{}
```

Figure 11.6 – The Notecard program's output

8. Our Notecard will connect to the cloud and synchronize for the first time. It would also publish its rough location to the Notehub interface, as shown in the following screenshot:

Figure 11.7 – Notecard connected to the Notehub interface

If you are struggling with the setup, we recommend reading through the following blog post: `https://bit.ly/3tvswyG`.

In the next section, we will test the keypad.

Testing the keypad

In this section, we will verify the SparkFun Qwiic Keypad. The keypad is interfaced to the Raspberry Pi Pico, as follows (the schematic is shown in *Figure 11.5*):

- GP9 → SCL
- GP8 → SDA
- 3.3V → VIN
- The GND pins tied together

Let's look at the code we need to test our keypad. The code sample we'll be discussing in this section can be downloaded from this book's GitHub repository as `code_keypad_test.py`. Follow these steps:

1. As always, the first step is importing the requisite modules:

```
import sys
from time import sleep
import board
import busio
import sparkfun_qwiickeypad
```

2. Then, we must initialize the I2C interface and the keypad:

```
i2c = busio.I2C(board.GP9, board.GP8)
keypad = sparkfun_qwiickeypad.Sparkfun_QwiicKeypad(i2c)
```

3. Now, let's check whether the keypad is connected and has been initialized correctly:

```
if keypad.connected:
    print("Keypad connected. Firmware: ", keypad.version)
else:
    print("Keypad does not appear to be connected. Please
check wiring.")
    sys.exit()
```

4. The keypad values are stored in a FIFO stack. We can retrieve the keypad values as follows:

```
button = -1
while True:
    # request a button
    keypad.update_fifo()
    button = keypad.button
    # Display the button value
    if button > 0:
        print("Button '" + chr(button) + "' was
pressed.")
    sleep(0.100)
```

5. Putting it all together, we have the following:

```
import sys
from time import sleep
import board
import busio
import sparkfun_qwiickeypad

i2c = busio.I2C(board.GP9, board.GP8)
# Create keypad object
keypad = sparkfun_qwiickeypad.Sparkfun_QwiicKeypad(i2c)

print("Qwiic Keypad Simple Test")

# Check if connected
if keypad.connected:
```

```
    print("Keypad connected. Firmware: ", keypad.version)
else:
    print("Keypad does not appear to be connected. Please
check wiring.")
    sys.exit()


print("Press any button on the keypad.")


# button value -1 is error/busy, 0 is no key pressed
button = -1


# while no key is pressed
while True:
    # request a button
    keypad.update_fifo()
    button = keypad.button
    # Display the button value
    if button > 0:
        print("Button '" + chr(button) + "' was
pressed.")
    # wait a bit before trying again
    sleep(0.100)
```

When we save the preceding code sample as code.py, we should see the following output:

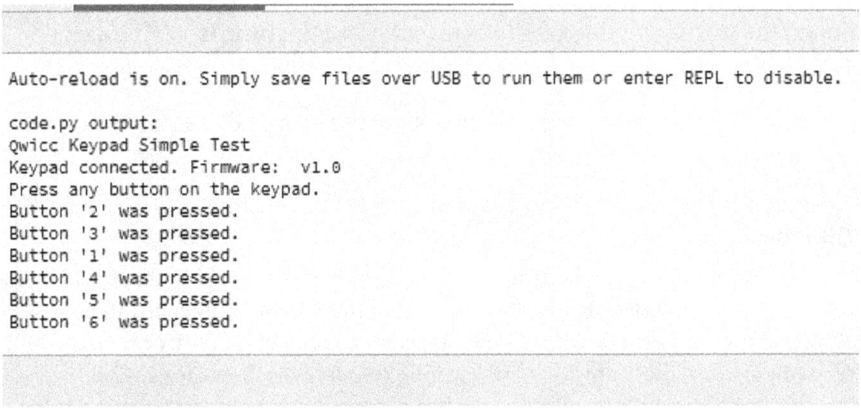

```
Auto-reload is on. Simply save files over USB to run them or enter REPL to disable.

code.py output:
Qwicc Keypad Simple Test
Keypad connected. Firmware:  v1.0
Press any button on the keypad.
Button '2' was pressed.
Button '3' was pressed.
Button '1' was pressed.
Button '4' was pressed.
Button '5' was pressed.
Button '6' was pressed.
```

Figure 11.8 – Keypad test output

Now that we've tested the individual components of our product, let's work on designing a PCB.

Designing a PCB

In this section, we will discuss designing a PCB. A PCB refers to a board where the connections between the components are routed using copper. These connections are called traces and the following figure shows the traces on a typical PCB:

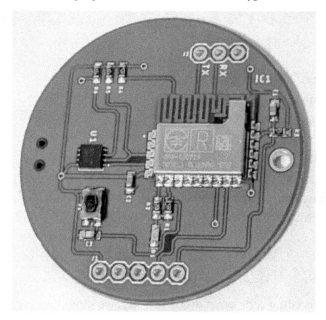

Figure 11.9 – Traces on a typical PCB

We recommend the learning guide on PCBs from SparkFun: `https://learn.sparkfun.com/tutorials/pcb-basics/all`.

In case you are not familiar with designing PCBs, we recommend the following course `https://teachmepcb.com/`.

PCB Design Software

There are various design tools available for designing PCBs. For this chapter's project, we used Autodesk Eagle. However, other tools are available, including KiCad, Altium, and others. KiCad is an open source project that is free to use. We used Eagle but there are plenty of learning resources available for KiCad.

Any PCB design has six major components to it, as follows:

- Schematic capture
- PCB layout
- Product enclosure considerations
- Gerber file generation
- Board fabrication
- PCB assembly

Let's discuss each in more detail.

Schematic capture

The schematic we will discuss in this section can be downloaded from this chapter's GitHub repository as blues-hardware.sch. During the schematic capture phase, we implement the schematic for our product. We make all the connections we discussed earlier in this chapter and these should look like what's shown in the following screenshot. The schematic can be downloaded from here, along with the PDF file: https://github.com/PacktPublishing/Raspberry-Pi-Pico-DIY-Workshop/tree/main/chapter_11/hardware.

Once the schematic capture is complete, an electrical rule check is run to ensure that there aren't any obvious errors, such as a missing connection.

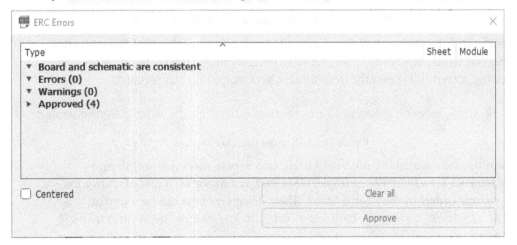

Figure 11.10 – Electrical rule check using Autodesk Eagle

Now that the schematic capture is complete, we will discuss the PCB layout.

PCB layout

The PCB layout we will discuss in this section can be downloaded from this chapter's GitHub repository as `blues-hardware.brd`. During the PCB layout phase, we route all the connections that were identified during the schematic capture phase. We also determine the shape of our PCB. The software helps with routing by actively tracking the number of connections left to route. Once the routing is complete, the completed board should look as follows:

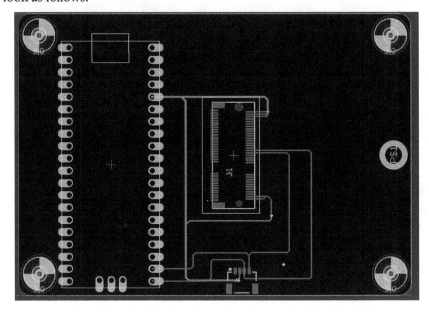

Figure 11.11 – Board routed with Autodesk Eagle

Once the PCB routing is complete, we need to run a design rule check to ensure that it does not violate any restrictions that have been set by the board manufacturer. The following screenshot shows the design rule check output for our product:

DRC: No errors. Left-click & drag to define group (or left-click to start defining a group polygon)

Figure 11.12 – Design rule check output

Design rules are available from board manufacturers to meet their minimum requirements for a PCB. For example, OSH Park is a major PCB manufacturer for low-volume orders in the United States. Their design rule file can be found at `https://docs.oshpark.com/design-tools/eagle/design-rules-files/`.

In the next section, we will discuss enclosure selection.

Enclosure selection

In our current design, our PCB shape is a rectangle (with dimensions of 83.9 mm x 58.2 mm) and we didn't necessarily consider enclosure. While designing a product, you need to keep enclosure requirements in mind. Most enclosures provide a recommended PCB template to follow. For example, the following is an enclosure manufacturer that provides PCB template files that can be directly imported into Eagle for design: `https://www.polycase.com/`.

In the next section, we are going to discuss Gerber file generation.

Gerber file generation

Once the PCB has been laid out, it is time to generate the Gerber files to send the PCBs out to a board house. Gerber is an open format that contains all the information that's needed to fabricate a PCB. The Gerber files for our product can be downloaded from this chapter's GitHub repository as `blues-hardware.zip`.

Gerber files are generated by scripts known as cam files. We used OSH Park's cam file to generate the Gerber (available from `https://docs.oshpark.com/design-tools/eagle/generating-gerbers/`).

Once the cam files have been generated, it is time to send the boards out for fabrication.

Board fabrication

The next step is board fabrication. There are several low PCB manufacturers. We have used OSH Park (`https://oshpark.com`), PCB Way (`https://pcbway.com`), and JLC PCB (`https://jlcpcb.com`). The process is fairly simple: you upload your Gerber files and pay for the fabrication costs. Once fabricated, they ship the PCBs to you. The PCB for our product looks like this:

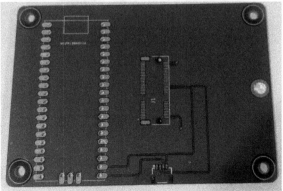

Figure 11.13 – Bare PCB upon arrival from fabrication

We have also uploaded our project to OSH Park and shared our project. It can be ordered from `https://oshpark.com/shared_projects/YKaQIo2P`.

In the next section, we will discuss assembling the PCB.

Assembling the PCB

Our product is a mix of through-hole and surface mount components. We soldered the M.2 connector using a reflow oven and soldered the Pico's headers by hand. Once assembled, the board should look like this:

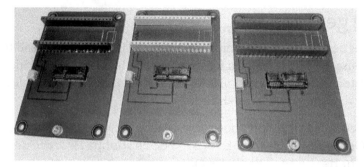

Figure 11.14 – Assembled PCBs

Once the board has been assembled, it is time to assemble the Raspberry Pi Pico and the Notecard, like so:

Figure 11.15 – Assembled board

Now that we have assembled the board, we will discuss bringing up the board for the first time.

Bringing up the board

In this section, we will discuss board bring-up. Board bring-up is a process where we verify the functionality of our design in a step-by-step process. It is identical to the steps we discussed earlier to test the Notecard and the SparkFun Qwiic keypad. During the verification process, we found a missing connection in our schematic. We connected it using a piece of wire (as shown in the following figure) and verified its functionality:

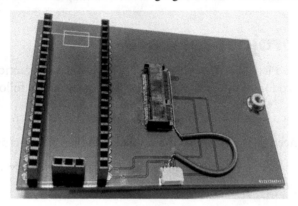

Figure 11.16 – Fixing a missing connection using a piece of wire

Now that we have tested our PCB, it is time to revise our design by adding the missing connection to the schematic.

Final assembly and testing

It is time to build our final product. We ordered an old landline phone with a keypad and gutted its internal circuitry. We glued our keypad and placed our PCB inside the phone. Our Pico phone looks like this:

Figure 11.17 – Assembled Pico phone

We recommend reading through our Hackster article to find out how we used it to text jokes when a person entered their phone number: `https://bit.ly/3pY7J5e`.

Here is a joke that our phone fetched from `icanhazdadjoke.com`:

Why did the barber win the race? He took a short cut.

That's a wrap! Let's summarize what we have built in this chapter.

Taking the project forward

So far, we have used the Pico as-is to develop our product. It is not practical to use the Pico development board for all designs. Some possible reasons are as follows:

- The size of the Pico board might not be suitable for your product.
- You might not need all the pins that are available on the Pico microcontroller.
- Compared to the Pico, an RP2040 microcontroller costs USD 1, but you do have additional components.

In the next section, we will discuss some options to replace the Raspberry Pi Pico in our design.

Replacing the Pico

To replace the Pico with a custom design based on the RP2040 microcontroller, we need to follow a reference design or an application note. Fortunately, the Raspberry Pi Foundation provides a document called Hardware design with RP2040 (available at `https://datasheets.raspberrypi.com/rp2040/hardware-design-with-rp2040.pdf`). The document details all the components that are needed for your design.

We will include a schematic that makes use of the RP2040 microcontroller instead of the Pico in this chapter's GitHub repository. There are two alternatives to making your design from scratch. The first one is called the PGA2040 and is available from Pimoroni.

Pimoroni PGA2040 – USD 9.05

The PGA2040 (shown in the following figure) is a breakout board that contains the minimal circuit needed to embed the RP2040 microcontroller in your design. The pins of the RP2040 microcontroller are made available in a grid. Pimoroni provides an Autodesk Eagle footprint to use in your designs. The PGA2040 can be purchased from `https://shop.pimoroni.com/products/pga2040`.

Figure 11.18 – PGA2040 board

The alternative to the PGA2040 is the RP2040 Stamp. We will discuss that in the next section.

RP2040 Stamp – USD 14.55

The RP2040 Stamp (shown in the following figure) is a breakout board that is about 1 square inch in size. Similar to the PGA2040, the Stamp exposes all the pins of the microcontroller. You can get the stamp working by providing a 5V supply to the board. The creator of the Stamp has provided footprints for both Autodesk Eagle and KiCAD. The RP2040 Stamp is available at `https://shop.pimoroni.com/products/rp2040-stamp`.

Figure 11.19 – RP2040 Stamp

The following figure shows the Stamp soldered onto a carrier board that comes in the Arduino form factor. This allows you to use the Stamp with Arduino shields:

Figure 11.20 – Stamp on a carrier board in the Arduino form factor

In this section, we have seen two alternatives that can be used instead of Pico. Now, let's summarize this chapter.

Summary

In this chapter, we discussed building a product using the Raspberry Pi Pico. We discussed building a proof-of-concept and verifying the individual components. Then, we performed schematic capture and PCB layout using Autodesk Eagle. This was followed by fabrication, assembly, and board bring-up.

We hope you enjoyed going behind the scenes and understanding the various steps involved in building a product. In the next chapter, we will discuss various tips and tricks you should keep in mind before we wrap up this book.

12
Best Practices for Working with the Pico

In this final chapter, we will cover some topics that weren't necessarily covered in the projects we discussed in this book. We will also discuss topics that will help you execute your projects better.

In this chapter, we will cover the following topics:

- Upgrading your Pico's firmware
- Programming the Pico using the Arduino IDE
- Programming in C/C++ using the Pico SDK
- Tools to aid with prototyping and product development

Technical requirements

There aren't strict hardware or software requirements for this chapter. You will need a Raspberry Pi Pico to discuss these topics. You can buy one from `https://www.adafruit.com/product/4883`.

Code in Action videos for this chapter can be viewed at `https://bit.ly/3kH2Cm7`.

We will start by understanding how to upgrade your Pico's firmware.

Upgrading your Pico's firmware

From time to time, you will need to upgrade your Pico's firmware. In this section, we will discuss upgrading your Pico's firmware. Upgrading your Pico's firmware might be necessary if you wish to make use of the latest version of CircuitPython or MicroPython on your Pico, which either includes new features in the Pico or fixes bugs that have been identified in the previous versions. The following steps are necessary to upgrade the firmware on your Pico:

1. We are going to assume that you are running either CircuitPython or MicroPython on your Pico. Let's discuss an example where we want to upgrade it to the latest version of CircuitPython. The first step is to download the latest version of the CircuitPython firmware that's available for the Raspberry Pi Pico from `https://circuitpython.org/board/raspberry_pi_pico/`.

2. Currently, the Pico that needs to be updated is running CircuitPython version 6.3.0. This information can be found from the CircuitPython interpreter running on your Pico, as shown in the following screenshot:

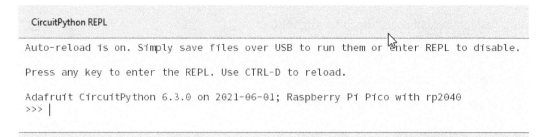

Figure 12.1 – CircuitPython firmware version shown in the REPL

At the time of writing this book, the latest version of CircuitPython is version 7.2.0. We have downloaded the latest firmware.

3. Now, we need to put the Pico into bootloader mode. Connect a USB cable to your computer, press and hold the BOOTSEL button (highlighted in the following figure), and connect a micro-USB cable to your Pico:

Figure 12.2 – The BOOTSEL button on your Pico

4. The Pico should enumerate as a storage device on your computer. The contents of your Pico should look similar to the following:

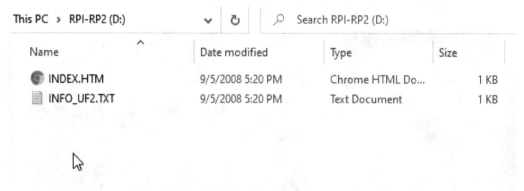

This PC › RPI-RP2 (D:)   ⌄   ↻   🔍 Search RPI-RP2 (D:)

Name	Date modified	Type	Size
🔴 INDEX.HTM	9/5/2008 5:20 PM	Chrome HTML Do...	1 KB
📄 INFO_UF2.TXT	9/5/2008 5:20 PM	Text Document	1 KB

Figure 12.3 – Contents of the Pico in bootloader mode

5. Now, copy over the downloaded firmware to the enumerated storage device.

6. Your Pico should reset and enumerate as a CircuitPython device, as shown in the following screenshot, where it enumerates as a storage device with the name CIRCUITPY:

› This PC › CIRCUITPY (D:) ›   ⌄   ↻   🔍 Search CIRCUITPY (D:)

Name	Date modified	Type	Size
.fseventsd	1/1/2020 12:00 AM	File folder	
lib	1/1/2020 12:00 AM	File folder	
.metadata_never_index	1/1/2020 12:00 AM	METADATA_NEVE...	0 KB
.Trashes	1/1/2020 12:00 AM	TRASHES File	0 KB
boot_out.txt	1/1/2020 12:00 AM	Text Document	1 KB
code.py	1/1/2020 12:00 AM	Python File	1 KB

Figure 12.4 – Pico enumerated as a storage device after the firmware update

7. Now, you can verify the firmware version by launching the CircuitPython interpreter in the Mu IDE.

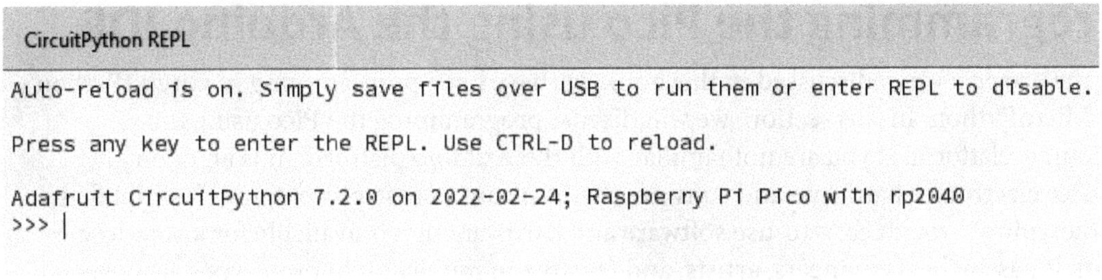

```
CircuitPython REPL

Auto-reload is on. Simply save files over USB to run them or enter REPL to disable.

Press any key to enter the REPL. Use CTRL-D to reload.

Adafruit CircuitPython 7.2.0 on 2022-02-24; Raspberry Pi Pico with rp2040
>>> |
```

Figure 12.5 – Latest firmware installed to the Pico

8. Alternatively, you can find the firmware version in the `boot_out.txt` file, which can be found on your Pico, as shown in the following screenshot:

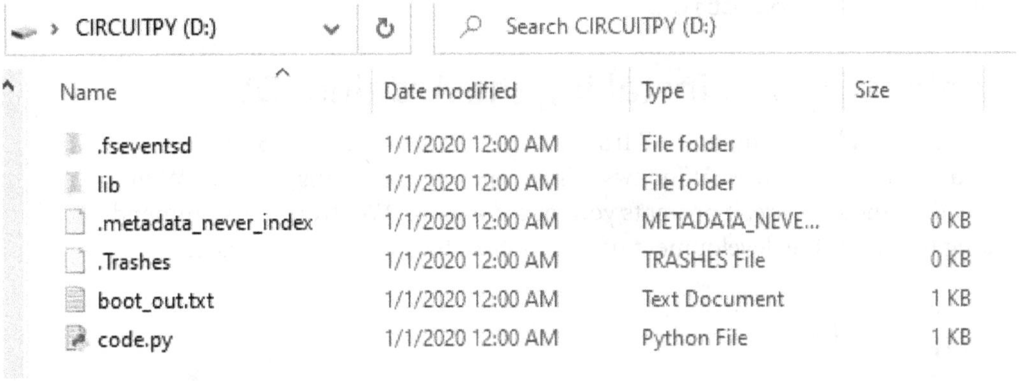

Name	Date modified	Type	Size
.fseventsd	1/1/2020 12:00 AM	File folder	
lib	1/1/2020 12:00 AM	File folder	
.metadata_never_index	1/1/2020 12:00 AM	METADATA_NEVE...	0 KB
.Trashes	1/1/2020 12:00 AM	TRASHES File	0 KB
boot_out.txt	1/1/2020 12:00 AM	Text Document	1 KB
code.py	1/1/2020 12:00 AM	Python File	1 KB

CIRCUITPY (D:)   Search CIRCUITPY (D:)

Figure 12.6 – boot_out.txt on the Pico enumerated as a storage device

9. The file contains the following information and is similar to the one that's found on the interpreter. As you may have noticed, this is the same information that's displayed on the CircuitPython interpreter:

```
Adafruit CircuitPython 7.2.0 on 2022-02-24; Raspberry Pi
Pico with rp2040
Board ID:raspberry_pi_pico
```

You can use the technique described in this chapter to either upgrade firmware or switch from CircuitPython to MicroPython and vice versa. We also recommend updating your Pico from time to time to make sure that you are running the latest firmware.

In the next section, we will discuss programming the Pico using the Arduino IDE.

Programming the Pico using the Arduino IDE

In the projects we've discussed in this book, we have been programming in CircuitPython or MicroPython. In this section, we will discuss programming the Pico using the Arduino platform. If you are not familiar with the Arduino platform, it is an open source electronics hardware and software that is meant to ease prototyping. It simplified prototyping through easy-to-use software and hardware that is available for a very low cost. It has enabled engineers, artists, and creators to rapidly build prototypes. There is a vast ecosystem of add-on hardware (known as **shields**) available for the Arduino platform. The Arduino platform enables the Pico to be programmed in C++. It also provides access to the vast ecosystem of libraries and add-on hardware that's available for the Arduino platform. You can learn more about the Arduino platform at `https://www.arduino.cc/en/Guide/Introduction`.

Downloading and installing the Arduino IDE

You can download the Arduino IDE from `https://www.arduino.cc/en/software`. It is available for Windows, Linux, and Mac operating systems. While downloading the software, it prompts you for a donation. We strongly recommend donating to further the development of open source hardware and software.

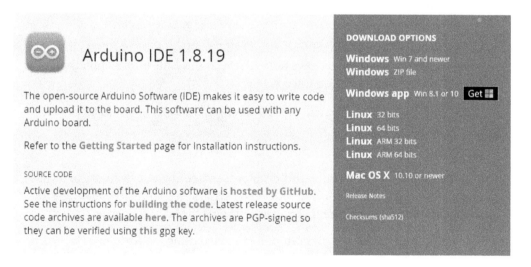

Figure 12.7 – Downloading the Arduino IDE

The installation process is straightforward. We will discuss installing the board support package for the Pico in the next section.

Installing a board package for the Pico

Now that we have installed the Arduino IDE, let's discuss installing the required packages that are meant to support the Raspberry Pi Pico in the Arduino platform. Let's get started:

1. Upon launching the Arduino IDE, it should launch a window, as shown in the following screenshot:

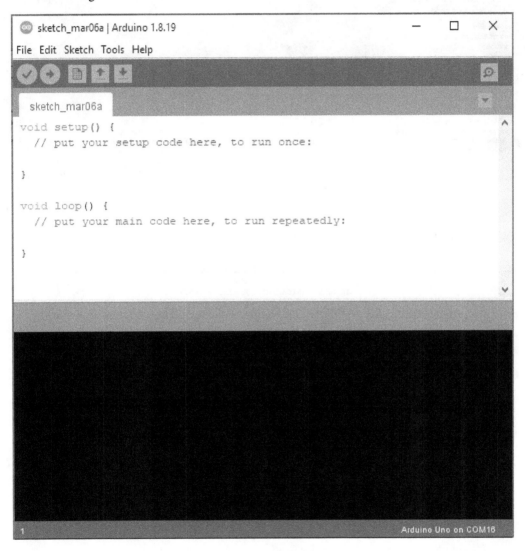

Figure 12.8 – Launching a new window of the Arduino IDE

2. Now, go to **Tools | Board | Boards Manager**. This should launch a new window containing all the board support packages that can be installed. From the search bar, look for packages that support the Pico, as shown in the following screenshot:

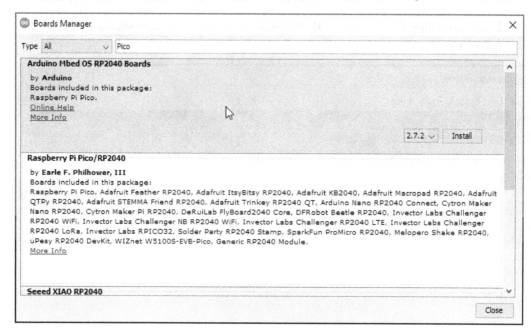

Figure 12.9 – Arduino Boards Manager

3. Install the **Arduino Mbed OS RP2040 Boards** package. This package includes support for the Raspberry Pi Pico.

4. To program the Pico, we need to select the necessary board. You can select the board by going to **Tools | Board | Arduino Mbed OS RP2040 Boards | Raspberry Pi Pico**, as shown in the following screenshot:

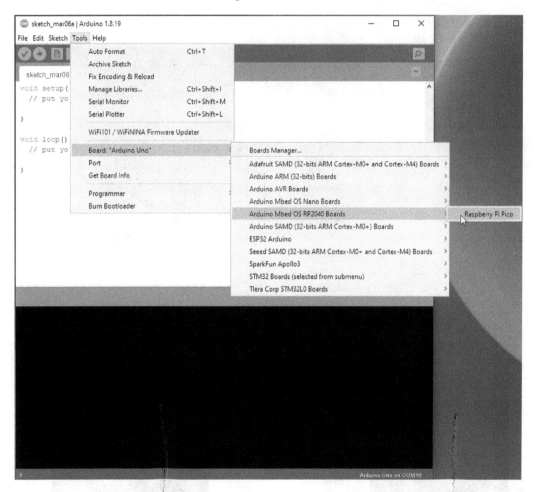

Figure 12.10 – Selecting the Raspberry Pi Pico

5. Now, it is time to load an example that's available from the Arduino IDE to blink the onboard LED that's on the Raspberry Pi Pico. Go to **File | Examples | Basics | Blink**, as shown in the following screenshot:

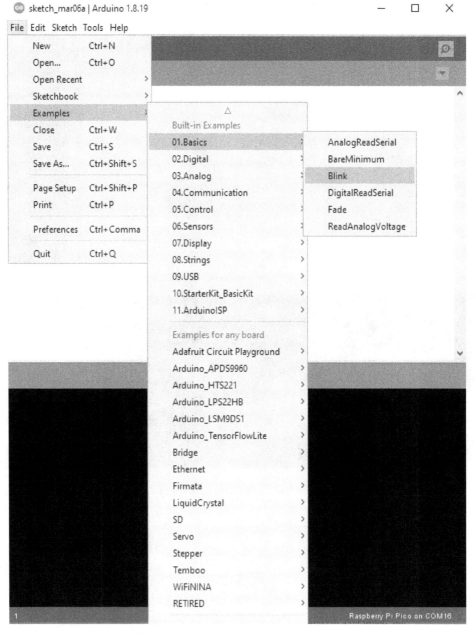

Figure 12.11 – Loading the Blink example

6. It should load the code sample, as shown in the following screenshot. Now, let's compile the code to verify whether there are any errors in our code. In this case, we are trying to verify whether the board support package installation works correctly. The *compile* button is a checkmark located at the top-left corner (highlighted with a red rectangle here):

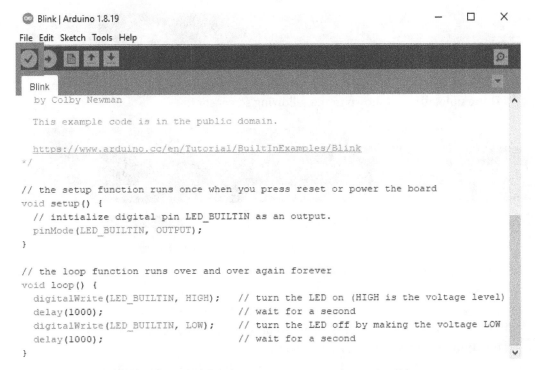

Figure 12.12 – Compiling the Blink example

7. Upon successful compilation, a message stating **Done compiling** will be shown at the bottom along, with the compiler's output, as shown in the following screenshot:

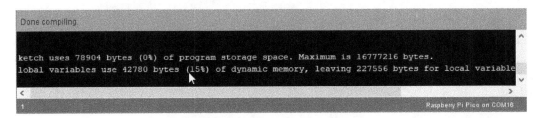

Figure 12.13 – Compiler output

8. Now, it is time to upload the code to the Pico. To upload the code to the Pico, you need to enter bootloader mode. We discussed putting the Pico in bootloader mode in the *Upgrading your Pico's firmware* section. Press the **BOOTSEL** button while connecting the micro-USB cable. Then, click the **upload** icon (the arrow button) next to the **compile** button (highlighted in the following screenshot with a red rectangle):

Figure 12.14 – Upload button

9. Once the compiled binary has been uploaded, it should display a message stating **Done uploading**, as shown in the following screenshot:

```
Done uploading.

Sketch uses 78904 bytes (0%) of program storage space. Maximum is 16777216 bytes.
Global variables use 42780 bytes (15%) of dynamic memory, leaving 227556 bytes for local variables
.

<                                                                                              >
1                                                                          Raspberry Pi Pico on COM17
```

Figure 12.15 – Compiled binary uploaded to the Pico

10. Once the code has been uploaded, the onboard green LED will start blinking in 1-second intervals.

11. Let's review the code sample that we just compiled and uploaded to the Pico. An Arduino code file is called a **sketch**. A typical Arduino sketch consists of two functions, namely, `setup()` and `loop()`.

12. The `setup()` function is where all the initialization for the program occurs. For the LED blinking example, we are setting the built-in LED connected to GPIO pin 25 as an output pin. In the following code snippet, `LED_BUILTIN` refers to GPIO pin 25. This is defined in the board support package for the Pico:

```
void setup() {
    pinMode(LED_BUILTIN, OUTPUT);
}
```

13. The `loop()` function is used to implement any code that needs to run inside an infinite loop. For this example, we need to blink an LED at a 1-second interval. To do so, we need to turn the LED on and off for 1 second between each function call. In the following code snippet, the `delay()` function accepts arguments in milliseconds. Hence, we have used `delay(1000)` to introduce a 1-second delay:

```
void loop() {
    digitalWrite(LED_BUILTIN, HIGH);
    delay(1000);
    digitalWrite(LED_BUILTIN, LOW);
    delay(1000);
}
```

Consider making the delay 100 ms to see the rapid blinking effect of the LED. Play with the interval to determine the smallest interval at which the blinking is visible to the naked eye.

Now that we've had fun with the Arduino IDE, we will discuss developing applications for the Raspberry Pi Pico using the official SDK from the Raspberry Pi foundation.

Programming in C/C++ using the Pico SDK

In this section, we will discuss developing applications using the official **software development kit (SDK)** from the Raspberry Pi foundation. The SDK allows you to develop applications in C/C++ and the Raspberry Pi foundation has released detailed documentation on the API to help you develop applications using the Pico. The documentation is available at `https://bit.ly/3IpBFwW`.

The documentation includes information on installing the SDK, compiling your first example, and loading it onto your Pico. We recommend working through the documentation if you are interested in programming in C/C++.

> **C/C++ Code Samples**
> We have included C/C++ code samples for the Raspberry Pi Pico in this book's GitHub repository.

Debuggers for the Raspberry Pi Pico

If you are learning to program in C/C++, it is useful to have a debugging tool to troubleshoot your code. The Raspberry Pi Pico comes with a **Serial Wire Debug** (**SWD**) port that lets you troubleshoot your code. It lets you step through your code line by line and observe the code's behavior, as well as the state of the variables. We recommend the following articles for setting up a debugger for application development:

- If you are using a Raspberry Pi for application development, the following article describes interfacing the Raspberry Pi Pico's SWD port to a Raspberry Pi: `https://bit.ly/3JwX8oT`.

- The following article describes setting up a Pico as a debugger (also knowns as a picoprobe) for another Pico: `https://bit.ly/3qkHt4i`.

In the next section, we will discuss some tools that can aid with prototyping and product development.

Tools to aid with prototyping and product development

In this section, we will discuss tools that can aid prototyping and your general product development needs. We will start by discussing a breadboard from Simon Monk.

Breadboard with Pico's pinout labels

In *Chapter 1, Getting Started with the Raspberry Pi Pico*, we discussed the Pico's pinouts. It can be cumbersome to keep referring to the pinouts while you are trying to build your prototypes. We came across a breadboard from Simon Monk (MonkMakes) that carries the labels on the Pico's pinouts (as shown in the following figure). This allows you to wire up your connections easily during prototyping.

In the following figure, the blue letters refer to the GPIO pin numbers, while the ground pin is indicated with a G and the power pins are indicated in red letters. The labeling scheme assumes that the Pico is mounted onto the breadboard, as shown in the following figure:

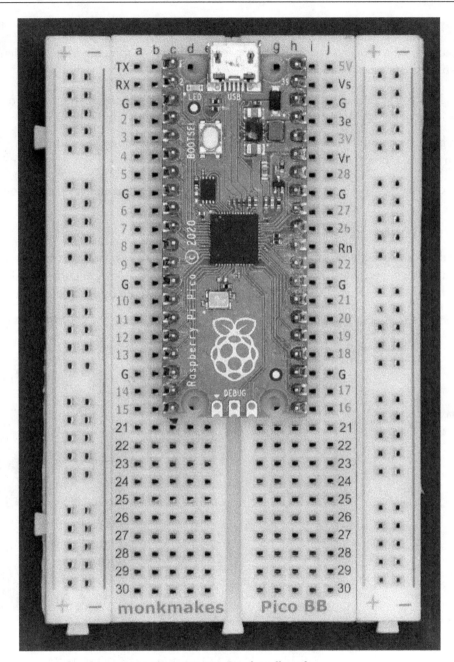

Figure 12.16 – Pico breadboard

In the next section, we will discuss power profiling your product to build battery-powered products.

Power profiling your application

In this section, we will discuss power profiling for product development. Power profilers serve two main purposes, as follows:

- To help you understand the overall power consumption pattern of your device or product. It helps you identify the battery/power source in your application.

- In battery-powered devices, a power profiler can help you understand the sleep current and identify any current leaks in your design.

We have come across two power profilers in two different price ranges. Let's take a look.

Nordic Power Profiler Kit – USD 81.25

This Nordic Power Profiler Kit from Nordic Semiconductors allows you to profile devices with input voltages between 1.8 V and 3.3 V and provides up to 70 mA of current. This power profiler is ideal for applications where the device is powered by a single-cell battery.

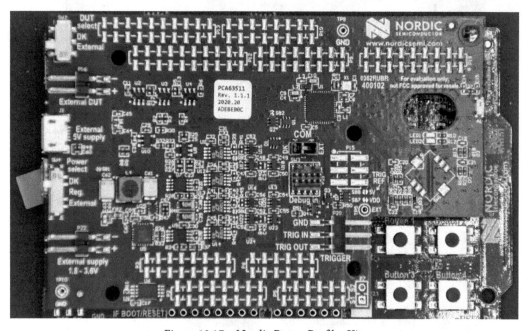

Figure 12.17 – Nordic Power Profiler Kit

The latest version of this kit is available for USD 92.50 at `https://bit.ly/3N6Rqwc`.

In the next section, we will discuss the Joulescope.

Joulescope – USD 999

The Joulescope is a rugged, expensive energy analyzer that can accommodate voltages between 1 and 15 V and up to 3 A of current with a 1.5 nA resolution. The Joulescope is expensive, but it is useful if you are developing products that have a wide range of current consumption. It is worth it if you are going to be pursuing a career as an electrical engineer.

Figure 12.18 – Joulescope for power profiling

In the next section, we will discuss programming the PIOs on the RP2040 microcontroller.

Programming the PIOs

PIO refers to **Programmable Input/Output**. It allows you to write assembly programs to implement features/interfaces that are generally not available on the RP2040 microcontroller. For example, the RP2040 microcontroller comes with two UART interfaces. Let's consider a scenario where you have used both interfaces and you need a third UART interface. You can implement one using the PIO. According to the datasheet, there are two PIO blocks with four state machines so that you can implement the following interfaces:

- The 8080 and 6800 parallel buses
- I2C
- 3-pin I2S
- SDIO
- SPI, DSPI, and QSPI
- UART
- DPI or VGA

PIOs allow you to run these interfaces independently of the main processor. The following diagram shows a single PIO block on the microcontroller:

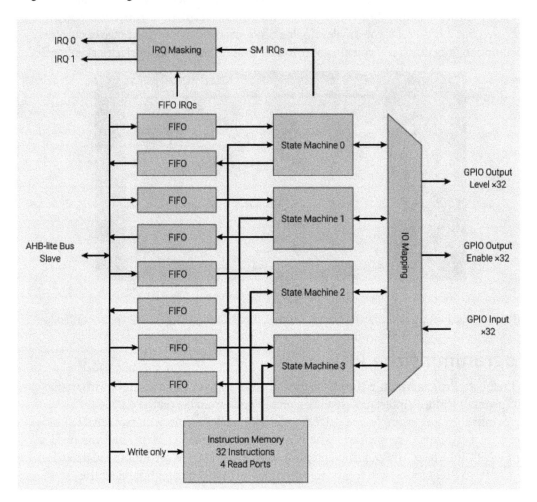

Figure 12.19 – Single PIO block (Image source: RP2040 datasheet)

Developers have interfaced a display into a VGA display using the PIO. We came across the following resources to help you make use of the PIO on the Raspberry Pi Pico:

- *Chapter 3* of the RP2040 microcontroller's datasheet: https://bit.ly/3wnG0ht

- Using CircuitPython to program the PIO: https://bit.ly/3N3Clvp

- Digikey's article on programming the PIO in C/C++: https://bit.ly/3qIcOyr

Now, it's time to wrap this book up with a final word.

Summary

And that's a wrap!

We hope you enjoyed reading the various projects we discussed in this book. We culminated this chapter with various tips and tricks we gathered while we were working on the other chapters in this book. We will maintain this book's GitHub repository with the latest code samples and updates as necessary. We recommend submitting an issue in this book's GitHub repository if you come across any bugs while working on a project that was discussed in this book.

We, as makers and citizen scientists, are glad you joined us on this journey in developing applications with the Raspberry Pi Pico. The maker movement has sprouted and grown to be a strong force across the world, touching people of all ages, races, and backgrounds. The citizen science movement is no different, touching on some of the same key principles of collaboration and openness with an urge to solve important personal or societal problems cost-effectively.

We also hope that beyond that, teachers, students, individual hobbyists, and entrepreneurs will also find something in these past pages that will allow for the expansion of knowledge, personal enrichment, and commercial solutions. These projects can also be extended to other architectures and devices. You may also be chasing cost or form factor efficiencies like us, and we hope that this book and the projects therein might be useful, whatever your objectives might be.

We hope you have fun making things with the Raspberry Pi Pico. Happy making!

Index

346

V

W

Packt.com

Subscribe to our online digital library for full access to over 7,000 books and videos, as well as industry leading tools to help you plan your personal development and advance your career. For more information, please visit our website.

Why subscribe?

- Spend less time learning and more time coding with practical eBooks and Videos from over 4,000 industry professionals

- Improve your learning with Skill Plans built especially for you

- Get a free eBook or video every month

- Fully searchable for easy access to vital information

- Copy and paste, print, and bookmark content

Did you know that Packt offers eBook versions of every book published, with PDF and ePub files available? You can upgrade to the eBook version at packt.com and as a print book customer, you are entitled to a discount on the eBook copy. Get in touch with us at customercare@packtpub.com for more details.

At www.packt.com, you can also read a collection of free technical articles, sign up for a range of free newsletters, and receive exclusive discounts and offers on Packt books and eBooks.

Other Books You May Enjoy

If you enjoyed this book, you may be interested in these other books by Packt:

Mastering ROS for Robotics Programming - Third Edition

Lentin Joseph, Jonathan Cacace

ISBN: 9781801071024

- Create a robot model with a 7-DOF robotic arm and a differential wheeled mobile robot
- Work with Gazebo, CoppeliaSim, and Webots robotic simulators
- Implement autonomous navigation in differential drive robots using SLAM and |AMCL packages
- Interact with and simulate aerial robots using ROS
- Explore ROS pluginlib, ROS nodelets, and Gazebo plugins
- Interface I/O boards such as Arduino, robot sensors, and high-end actuators
- Simulate and perform motion planning for an ABB robot and a universal arm using ROS-Industrial
- Work with the motion planning features of a 7-DOF arm using MoveIt

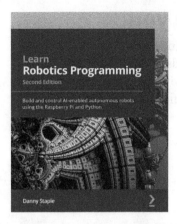

Learn Robotics Programming - Second Edition

Danny Staple

ISBN: 9781839218804

- Leverage the features of the Raspberry Pi OS
- Discover how to configure a Raspberry Pi to build an AI-enabled robot
- Interface motors and sensors with a Raspberry Pi
- Code your robot to develop engaging and intelligent robot behavior
- Explore AI behavior such as speech recognition and visual processing
- Find out how you can control AI robots with a mobile phone over Wi-Fi
- Understand how to choose the right parts and assemble your robot

Packt is searching for authors like you

If you're interested in becoming an author for Packt, please visit authors. packtpub.com and apply today. We have worked with thousands of developers and tech professionals, just like you, to help them share their insight with the global tech community. You can make a general application, apply for a specific hot topic that we are recruiting an author for, or submit your own idea.

Share your thoughts

Now you've finished *Raspberry Pi Pico DIY Workshop*, we'd love to hear your thoughts! Scan the QR code below to go straight to the Amazon review page for this book and share your feedback or leave a review on the site that you purchased it from.

https://packt.link/r/1801814813

Your review is important to us and the tech community and will help us make sure we're delivering excellent quality content.